The Electron

A Biography

Brian Clegg

ICON

Published in the UK in 2026 by
Icon Books Ltd, Omnibus Business Centre,
39–41 North Road, London N7 9DP
email: info@iconbooks.com
www.iconbooks.com

ISBN: 978-183773-332-3
ebook: 978-183773-333-0

Typesetting by SJmagic DESIGN SERVICES, India

Printed and bound in the UK

Appointed GPSR EU Representative: Easy Access System Europe Oü, 16879218
Address: Mustamäe tee 50, 10621, Tallinn, Estonia
Contact Details: gpsr.requests@easproject.com, +358 40 500 3575

For Gillian, Rebecca and Chelsea

ACKNOWLEDGEMENTS

Writing a biography inevitably increases your interest in, and affection for, the subject. I have found this when dealing with subjects such as Roger Bacon, Eadweard Muybridge and James Clerk Maxwell, and the same has been true of the electron. Although I knew intellectually how important the electron is, it was only by getting to know it better that I have come to appreciate its significance at a gut level.

Many people have contributed to this journey – my physics teachers and lecturers (including Oliver Ridge, who allowed school friend David Ball and me to attempt to construct a laser from scratch in our final months at the Manchester Grammar School) and the many science writers whose work I have enjoyed over the years, including those in the Further Reading section at the end of this book. At Icon Books, Connor Stait, Nikki Embery and Steve Burdett, who have helped shape the contents.

CONTENTS

I SING THE BODY ELECTRIC 1

I sing the body electric,
The armies of those I love engirth me and I engirth them,
They will not let me off till I go with them, respond to them,
And discorrupt them, and charge them full with the charge of the soul.

Walt Whitman, 'I Sing the Body Electric'
(this version 1867)

According to Google, the most significant thing to spring to mind when mentioning 'electron' is 'a free and open-source software framework' of that name. Wonderful though this framework may be, the real electron is far more important. It's a fundamental particle. The driving force of electricity and electronics. The enabler of chemistry and life. An essential part of stuff. There may be a degree of dispute over exactly what Walt Whitman had in mind when he came up with that wonderful phrase 'I sing the body electric'* – but there is no doubt that our bodies are entirely dependent on the electron's presence.

---

* Out of interest I asked a few AI chatbots what they thought 'I sing the body electric' meant: the general feeling was a combination of vitality and the linking of the body and the soul. As we will see later, the connection made between electricity and some form of 'life force' would be a strong one. For Ray Bradbury fans, though, it is more likely to bring to mind one of his short story collections.

In attempting to uncover the electron's biographical details we need to go back a long way. We sometimes get excited about old things. But what counts as old? Human constructions such as Stonehenge and the pyramids, built thousands of years ago? Rocks that date back as much as 4 billion years? These are far older than the fleeting lifetime of a human being. Yet every atom in your body contains electrons that have been in existence since shortly after the Big Bang, 13.8 billion years ago. They are tiny, negatively charged and fundamentally important for existence. Ever since the creation of the universe (however you want to define creation), electrons have been there, singing the song of their electric existence.

Any good biography will concern itself with both life and death. Left to their own devices, electrons are to all intents and purposes immortal. The best estimate to date suggests that the particles should survive around 6.6 x 10^{28} years (10^{28} is 1 followed by 28 zeroes), around 5 x 10^{18} times as long as they have so far been around. Coming up with a figure for something expected to occur so far into the future is not trivial. The value given is based on an experiment using the Borexino detector at the Gran Sasso National Laboratory, located below a mountain in Italy to reduce impact from cosmic rays and other incoming particles.

The logic for that ludicrously large lifetime is based on the vast number of electrons present in the detector (around 10^{32}), none of which were observed to decay during the run of the experiment. In theory, an electron could decay into either three neutrino particles or a neutrino and a photon, but this has not been observed. If you have dipped a toe into particle physics you might think that an electron decaying into three neutrinos or a neutrino and a photon makes no sense, as its electrical charge would have to disappear, but theoreticians suggest that an electron can only decay if it violates charge conservation.

The reason for this is that the electron is the lightest particle with an electrical charge. If it broke up, all the available lighter particles it could produce would have no charge. The only alternative would be if it decayed into a totally new charged particle – but if this existed, it is hard to understand why it hasn't already been seen. Back in 2015, Gianpaolo Bellini, working with the Borexino detector, commented that with better elimination of background radiation, they might be able to boost the estimated minimum lifetime to 10^{31} years.

Although this astonishing lifespan makes electrons sufficiently close to immortal to be far beyond any human considerations, this doesn't mean that electrons can't die far sooner, not by breaking into pieces but by being totally destroyed. As we will discover later, the electron has a nemesis known as a positron – an antimatter electron. When the matter and antimatter versions are brought together, they annihilate, turning from matter into pure energy in the form of photons: particles of light. The reverse can also happen, with photons producing matter–antimatter pairs (this is where electrons came from in the first place), but for an individual electron, an encounter with its nemesis brings its life to an end.

In the beginning

According to our best current theory of the origins of the universe, things started small with a rapidly expanding space, filled with what was initially pure energy in the form of photons. 'Let there be light' got it right. But, as Einstein showed, in the right circumstances energy and matter are interchangeable, and the initial energy of these particles was so high that the photons readily transformed into matter. A pair of photons could easily end up as an electron and its antimatter equivalent, the positron.*

The energy produced when electrons and positrons annihilate can sound distinctly insignificant. A single pair produces a tiny amount of energy: it would take around 100,000 billion (10^{14}) pairs to power a typical LED bulb for a second. But a combination of matter and antimatter is also the most compact way to store energy. If you had just a gram of electrons and a gram of positrons it would produce around 1.8×10^{14} joules – that's around the entire electricity consumption of the UK for three minutes.

---

* A single photon can also produce an electron–positron pair, but this usually requires the existence of an atomic nucleus to satisfy the conservation of momentum, and these would not have existed when electrons first formed. If two photons are involved, the momentum issue can be solved without an additional nucleus.

The pairs of matter particles produced in the early universe would have been constantly recombining in a dance of production and elimination. But by around fifteen minutes after the Big Bang, as the universe expanded, the energy of the photons would have dropped to the extent that they were no longer producing new matter at any scale.

Now we get a bit of a problem that is yet to be resolved. Unless they were somehow separated, the electrons and positrons should have continued annihilating each other, leaving no matter behind. Some theories suggest there might have been a little more matter than antimatter, resulting in around one in a billion electrons surviving, though it's not really clear that this would produce what we now experience. Somehow, though, we did end up with an observable universe that is almost entirely made up of matter rather than antimatter.

In the subsequent seconds, other fundamental particles called quarks, which had also been produced from energy in a so-called 'quark-gluon plasma' (gluons are the particles that are able to bind quarks together), started to combine to produce the familiar particles of the atomic nucleus: protons and neutrons. Again, a small excess of particles over antiparticles seems to have resulted in a conveniently identical number of positively charged protons to the number of electrons, leaving the universe electrically neutral.

At this point, though, we didn't have neutral atoms. Instead, the electrons were free in space while hydrogen nuclei (some of which underwent nuclear fusion to produce helium nuclei, much as now happens in stars) were positively charged. It was only after about 370,000 years that energy levels had dropped sufficiently with the continuing expansion of the universe for the electrons to become attached to the positively charged nuclei. Without all the charged particles floating around, which prevent light from moving unhindered, photons were able to start flying freely through the universe, producing radiation we can still detect, known as the cosmic microwave background.

If we need to look at familial origins for a good biography, the vast majority of the electrons we now encounter came into being in those opening minutes of the universe's existence. Each one is around 13.8 billion years old. It's a very flat family tree, with the majority of occupants all cropping up on the same line. But ever since heavier elements were formed, there has been a potential for new electrons to come into being, right up to the present day. These are the electrons responsible for beta radiation.

When an atomic nucleus decays in the process often known as nuclear fission, a neutron can be transformed into a proton, an electron and a third particle called an antineutrino. This electron is newly formed, rather than one that originated just after the Big Bang. But these are statistically unnoticeably few compared with the vast majority that have been in existence since the origin of the universe.

Strength in numbers

An individual electron has very little mass or electrical charge – but their strength is in numbers. According to physicist Frank Close, 'with each breath, you inhale about ten billion trillion atoms of oxygen, each containing several negatively charged electrons. In all, this amounts to about 15,000 coulombs of electricity, enough to spark 1,000 bolts of lightning ... If you are breathing steadily the intake corresponds to an electric current of some 3,000 amps ...'

To put that into context, the familiar unit of amps measures the flow of electrical current. In the UK, standard mains sockets are rated at 13 amps. The fastest electric car chargers at the time of writing can run at up to around 1,000 amps. A coulomb measures charge – effectively how many electrons are involved. A 1-amp current for a second provides a charge of 1 coulomb – which amounts to around 6×10^{18} electrons.

Of course, we don't experience this (literally) shocking electrical experience when we breathe because the negative electrical charge of the electrons is balanced out by the positive electrical charge of the protons in the nuclei of the atoms we are breathing. By a happy coincidence, the negative electrical charge on the electron is exactly the same as the positive electrical charge on the proton. This feels even more remarkable when you realise that a proton is made up of three smaller particles called quarks. Two are up quarks, with a positive charge two-thirds that of the electron's negative charge, and one is a down charge with a one-third negative charge.

A 'happy coincidence' doesn't feel like a good scientific explanation, but we don't know why this is the case, except that matter as we know it would not exist if it were not so. There is a quasi-scientific argument that charge is quantised (i.e. comes in minimum sized buckets) – but this doesn't actually explain in any sense why the electrical charge should be conveniently balanced in this way.

Without this cancelling out of charge, gravity would seem trivial. Gravity is much, much weaker than electromagnetism, the fundamental force behind the interaction of electrical charges. Rub an inflated balloon on your hair and it will gain an electrical charge. The rubbing action transfers electrons from your hair to the balloon's surface. Bring the negatively charged balloon near to small pieces of paper and it will pick them up. The small electrical charge on the balloon has a stronger attraction than the entire gravitational pull of the massive Earth.

Because two negative charges repel each other, electrons are naturally antisocial. If you imagine trying to get two electrons closer and closer, the force pushing them apart grows as the square of the distance between them falls, just as the gravitational attraction between their masses grows with the square of the distance. But that repulsive electrical force is around 4×10^{42} times stronger than the attraction of gravity.

For the vast majority of the lifetime of electrons, they have been going about their business in reactions (or simply sitting around in atoms) without having any interaction with human beings. It would make for a dull biography if we spent any time on the intervening 13.8 billion years in which very little happened in terms of understanding electrons. But the story changes with the emergence of human beings. The earliest hominins – distantly related species – date back around 7 million years, with the smaller brained *Homo* species originating between a million and 1.5 million years ago. Exactly when our own species, *Homo sapiens*, emerged isn't clear, but estimates range from around 190,000 to 300,000 BCE.

However, there is no opportunity to develop the electron's biography in any detail until we get to written history. For that reason, the rest of the book will primarily stick to the last few hundred years when we have started to consider electrons and their most immediately obvious role as the carriers of electrical current.

Whitman was right to refer to the body electric, though the science of his day could not fully back up the connection. Galvani had demonstrated a relationship between electricity and the twitching of muscles. Speculation had been rife for centuries about the relationship between electricity and life – could it be the vital spark, the 'spirit' as Newton called it, that distinguished the living from the dead? Perhaps the most vivid picture we might have is Mary Shelley's character Victor Frankenstein giving life to his creature using a bolt of electricity.

It's a dramatic image, though sadly it's not a picture that emerges from Shelley's imagination.* Frankenstein's deployment of the electrical power of the heavens in his laboratory was added to the story for the 1931 film. But it's no surprise that its director James Whale would depict a stream of electricity from the sky. Lightning must surely predate every other way that humans have consciously experienced the song of the electron in action.

Although gaining any understanding of lightning has required a scientific viewpoint, we have surely been aware of its appearance and its potentially devastating power for as long as *Homo sapiens* and our ancestors had conscious awareness of their surroundings. Few other natural phenomena make such an impact on our senses.

* Strictly speaking I should say that the story of Frankenstein emerged from Godwin's imagination, as Mary Shelley was not yet married when she came up with the first version of her story.

BENJAMIN FRANKLIN AND THE KITE OF DOOM

2

To demonstrate, in the completest manner possible, the sameness of the electric fluid with the matter of lightning, Dr Franklin, astonishing as it must have appeared, contrived actually to bring lightning from the heavens, by means of an electrical kite, which he raised when a storm of thunder was perceived to be coming on.

Joseph Priestley, *History and Present State of Electricity* (1767)

For many of us, looking out of the window at an electrical storm provides a fascinating and dramatic light show – and for pretty well all of us, being outside in an exposed location when there's lightning about is terrifying. So dramatic is the whole business of thunder and lightning that it was often allocated to the gods – Zeus, Jupiter, Indra and Thor all had thunderbolts at their disposal. Even without a religious connotation, Pliny the Elder, writing in the first century CE, considered lightning prophetic, direful and accursed. This was the electron in the wild – electricity rendered visible and awesome.

Back then, the associated rumble of thunder would have been considered something other than the direct sound of a lightning bolt in action, because we usually hear the thunder considerably later than we see the flash. This is because of the difference in speed between light at 300,000 kilometres (180,000 miles) per second and sound at 340 metres (1,115 feet) per second. If a lightning bolt is 10 kilometres (6.2 miles) away, the thunder will turn up around 29 seconds after we see the flash. But what we are hearing is the action of an immensely powerful bolt of electricity ripping through the air.

This flood of charge zaps the air temperature up to as much as 30,000 °C (54,000 °F) – around six times the surface temperature of the Sun. The sudden heating accelerates air molecules so much that the expanding shock wave produces the equivalent of a sonic boom, though one that will be extended in time when heard because of the different routes that the sound waves can take to reach us.

That kite

Although it may have faded in modern experience in much the same way that many of the 'classic' novels from the past have become unfamiliar, a one-time instant association with attempting to understand the nature of lightning would have been Benjamin Franklin and his kite. This wasn't, of course, the first attempt to understand any aspect of electricity, but because the electron made its dramatic presence felt in lightning long before any other electrical phenomena were familiar, it seems an appropriate first step in uncovering the particle's biographical past.

The quote at the opening of the chapter comes from English scientist Joseph Priestley's attempt to summarise what was known about electricity in the mid-eighteenth century, and it was one of the early ways that Franklin's alleged experiment was spread far and wide. I say 'alleged' because, while it's clear that Franklin came up with the concept behind the risky undertaking, it is not by any means certain that he took the risk himself.

Throughout this book we will bring into the electron's biography individuals whose work has brought us closer to the electron and our knowledge of its role. Along the way, we will discover that a surprisingly high percentage of winners of the Nobel Prize in Physics have been working on the electron or on related phenomena. But this honour would not be introduced until 1901, when German physicist Wilhelm Röntgen won the first of these annual awards for his work on X-rays (which were generated from the energy of electrons). For the first few chapters, our central characters could not receive such an accolade, though arguably many would have done so, had the prize existed.

Franklin and the kite.

Alfred Jones, for the Bureau of Engraving and Printing, Public domain, via Wikimedia Commons

Benjamin Franklin was a superstar of the eighteenth century. Born in Boston in 1706, he initially made his money editing the *Pennsylvania Gazette* and publishing *Poor Richard's Almanack,* a popular annual mix of household guidance and amusement. A natural philosopher with a particular interest in electricity, and with good contacts in scientific bodies in the UK, he was also a diplomat and politician, classed as one of the Founding Fathers of the US from his position on the five-person committee that wrote the US Declaration of Independence.

The once-familiar story has Franklin venturing out at personal risk in a storm in 1752, flying a kite towards looming thunderheads. Despite the way it used to be popularly portrayed, this experiment did not involve trying to get the lightning to strike the kite, but rather for it to pick up an electrical charge from the clouds that was strong enough to send a stream of electricity through the air down to a metal key attached to the string. The key was linked by wire to a Leyden jar, a crude storage mechanism for static electricity. The point of the experiment was to demonstrate that the dramatic aerial phenomenon involved the same kind of electrical discharge (described by Franklin as 'electrical fire') possible on a much smaller scale from the electrical experiments that we will discover in the next chapter.

It's certainly true that Franklin wrote an article in 1750 proposing such an experiment, and some believe that Priestley's account corroborates the existence of an actual experiment, making it worthy of further examination. The doubt arises from the way that Franklin, never backward at coming forward, only wrote of the experiment indirectly. By the time his experiment is said to have taken place, something similar had already been carried out by the French scientist Thomas-François d'Alibard, inspired by Franklin's article. Franklin then wrote of a local attempt, 'the same Experiment has succeeded in Philadelphia, tho' made in a different and more easy Manner ...' But there is no mention of his own involvement.

Priestley, better known for discovering oxygen, was an irregular attender of the Lunar Society of Birmingham, a gathering of the great and good of the English Midlands who collectively had an interest in science (more on the society in chapter five). So it was Benjamin Franklin himself, who met Priestley when visiting England, who would provide Priestley with the information that featured in his book. We can't take Priestley's write up as absolute confirmation that Franklin's experiment took place, but he describes plenty of detail. The first observation was strands of the hemp kite string rising, as a build-up of electrical charge caused repulsion between charged strands. Franklin is then said to have brought his knuckle near the key and received a visible spark, before transmitting some electricity to a Leyden jar.

To complete this chapter, we need to move on to the better understanding of lightning that we now have. Most biographies proceed in a linear fashion through time. But as the electron has many different roles, it makes sense to follow each of these roles individually through time. The chapters are arranged in order of discovery – and it seems reasonable to assume that lightning was the first specifically electrical phenomenon that was known about – but each chapter might also venture into the future, to see how understanding has grown, as we are about to do.

There's the rub

Perhaps surprisingly, it still isn't totally clear exactly how the huge electrical charge needed for a lightning bolt builds up in the clouds. It seems certain, though, that this is a triboelectric effect. The 'tribo' part here comes from tribos, the ancient Greek word for rubbing – this is electricity generated by a rubbing motion.

As we saw in the previous chapter, a familiar example of triboelectricity is rubbing a balloon on your hair. This results in electrons being rubbed off the hair to give the balloon an overall negative electrical charge. In the best supported theory for the production of lighting, particles of ice and supercooled water droplets in the thundercloud are jostled against each other, transferring electrons in the same way. The heavier ice particles, carrying electrons, fall lower in the cloud, accumulating many small charges to form a powerful negatively charged region in the cloud.

For what happens next, it's best to look back at the way a charged balloon attracts small pieces of paper to it. Bring a charged object near to something that has no overall charge and it will have an influence. Because the balloon is negatively charged, it will repel the nearest electrons on the pieces of paper, forcing the electrons away.

Let's take a moment to remember what most of us learned at school. Two bodies with the same electrical charge repel each other, while opposite charges attract. It would be nice to be able to explain why this is the case – physicists may tell you it's about particle spin and quantum fields and relativity, but to be honest, this just pushes back the question, 'Why?' one level. All we can really say is that this is what electrical charges do.

As a result of the balloon pushing away nearby electrons in the paper as it is brought close, the side of the paper nearest the balloon will have lost some of its negativity and become positively charged. And because opposite charges attract, the paper is attracted to the balloon. The process of establishing a charge remotely this way is known as induction.

Move to the lightning picture and we have a huge negative charge towards the bottom of the thundercloud. This induces a positive charge on the bits of the Earth below that are closest to it – or, even more likely, in other nearby clouds. Around 80 per cent of lightning bolts travel from cloud to cloud rather than ever striking the ground. The voltage is sufficient to produce a weak electrical current, known as a leader, through the damp air between cloud and ground. This current results in some of the atoms between cloud and ground becoming ionised.

Ions are simply atoms that have gained or lost one or more electrons to become electrically charged in their own right. And ions are excellent electrical conductors. With a pathway now formed, the main, visible stroke can occur. This so-called return stroke actually travels from Earth to cloud. This leader/return stroke mechanism means that it's possible to generate a lightning bolt by firing rockets with a wire attached into a thundercloud. The wire provides the leader, resulting in a controlled discharge.

The power of nature

Though extremely brief, the power carried by a lightning strike can be immense. It's worth having a quick aside on energy and power, as these words will feature regularly. We tend to use the terms energy and power in everyday speech in a hand-waving fashion that is not the same as their use in physics. Energy is the ability to make things happen, measured by scientists in joules.* Work, an associated term, involves transferring energy from place to place. And power, measured in watts, is the rate at which energy is produced or used to do work. A single watt is a joule per second, though for domestic use it is often measured in kilowatts.

---

* Food energy is often measured in calories. Somewhat confusingly, what is often called a calorie is actually a kilocalorie – a thousand calories. A kilocalorie is 4,184 joules.

It's a scientific truism that energy is conserved. It can't be created or destroyed. Which makes the statement in the previous paragraph that power is the rate at which energy is produced or used distinctly confusing language. For convenience, we speak of a power station producing energy, or us using up energy when we are active. But really, we mean that a power station converts one kind of energy (say heat) into another (electrical energy). And when we run, we convert chemical energy into other forms, such as the kinetic energy of motion, potential energy if we're going uphill, and heat.

Transferring this knowledge of energy and power back to a lightning strike, on average a single bolt is roughly equivalent to the output of a mid-sized power station running for a second. That amounts to around half a billion joules, where 20 joules is what's required to run a 20-watt LED bulb for a second. A kilowatt hour, the measure typically used for consumption of household and electric vehicle energy, is around 3.6 million joules – so each bolt carries around 140 kilowatt hours. That's more than the entire storage of a high-capacity electric vehicle, typically expended in less than a second.

We can't see moving electrons per se – that's not what makes up the glowing, forked lightning bolt that splits the sky. Atoms give off light when the electrons in their structure go from a higher to a lower level of energy. The energy is lost from the atom in the form of a photon of light. The electrical discharge pushes the electrons in the atoms that make up the air and water through which it travels to a higher energy level. Those electrons then fall back in energy, forming the flash. This isn't all visible light, incidentally. Photons come in a whole range of energies.

This span is generally known as the electromagnetic spectrum, running from relatively low energy radio photons, through microwaves, infrared, visible light, ultraviolet, X-rays and gamma rays. Usually, lightning won't have sufficient energy to produce the most dangerous end of the spectrum, gamma rays. However, around 0.1 per cent of thunderstorms do have sufficient energy to generate these potentially deadly photons. This process is so energetic that pairs of electrons and their antimatter equivalent, positrons, can be produced. These will usually recombine, annihilating to generate more gamma rays.

Such lightning is relatively hard to see as there is little visible light given off. If a plane were to pass nearby, a single gamma-ray-producing bolt could have the same radiation impact as around a hundred chest X-rays, though thankfully commercial pilots do not go near large thunderclouds because of their potential impact on the safety of the flight.

We tend to think of thunder and lightning as a relatively rare phenomenon, because we may only witness it once or twice a year. But at any one time there are likely to be around 1,800 thunderstorms playing out around the world. According to myth, lightning doesn't strike the same location twice – but scientifically this is clearly nonsense. Lightning has no memory, so how could it possibly know where to avoid? Perhaps the most famous lightning target is the Empire State Building in New York, which is hit regularly, with as many as fifteen strikes from a single storm. Less numerically impressive but a more dramatically satisfying target is US park ranger Roy Sullivan, who survived seven separate lightning strikes.

One thing that Franklin definitely did do on the 'lightning = electricity' front was to invent the lightning conductor or lightning rod. These are pointed or forked pieces of metal attached to the highest point of a tall building, with a thick metal connector that runs down to the Earth. Without doubt this was a better idea than a medieval conception of lighting-strike prevention by putting a so-called thunderstone in the chimney of your house.

Based on that mythical concept of not striking twice, a thunderstone was a stone with a hole in it, which, according to folklore, had been caused by a lightning strike. In practice these stones seem to have been Stone Age hammer heads or later worked stones. The lightning conductor has more science behind it, though there has always been a dispute as to whether it was supposed to take a strike and conduct the electricity harmlessly to earth, or to make it less likely that a strike would happen in the first place, as it would start conducting charge away as it built up, before the charge reached the point of producing a leader. Even now there is dispute over whether pointed or rounded ends work better on a lightning conductor, and limited evidence on how much protection they offer.

Lightning will indubitably have given humanity its first experience of electricity, even though it would not be until Franklin's time that it was definitively identified as an electrical phenomenon. But for more than 2,000 years humans have been familiar with the smaller-scale production of electricity that gave the phenomenon (and indirectly the electron) its name. We next have to take a step back in time to encounter electrical bodies.

THESE SEVERAL BODIES (ELECTRICS) 3

These several bodies (electrics) not only draw to themselves straw and chaff, but all metals, wood, leaves, stones, earths, even water and oil.

William Gilbert (Trans. Edward Wright),
De Magnete Magneticisque Corporibus, et de Magno Magnete Tellure
(On the Loadstone and Magnetic Bodies and the
Great Magnet the Earth) (1600)

How did the 'electron' get its name? It's frequently said to be taken from the ancient Greek word for amber – but the reality is a little more complex than that. The substance amber has fascinated people for millennia. It's a little like a stone, but it is both too light and too soft. It is transparent yet is quite different from a crystal. And it can have once living things embedded in it. Particularly relevant, amber appeared to have special, magnet-like properties – when rubbed it could make small objects move towards it.

It's certainly true that the word 'electricity' was amber-inspired. In Latin, amber was called electrum, while in ancient Greek it was elektron, which derived from elector, roughly translatable as 'beaming Sun', as myth had it that amber was the tears of the daughters of Helios, the Sun god, remembering their dead brother. When first used, 'electricity' referred to the property of amber to attract materials when rubbed. Sir Thomas Browne in his *Pseudodoxia Epidemica* from 1646 notes 'Crystal will calefy* into electricity; that is, a power to attract strawes or light bodies, and convert the needle freely placed.' Browne was picking up on William Gilbert's Latin term 'electricus' to describe the nature of substances like amber to produce electrical attraction, something of an aside in his book on magnetism.

When it came to the naming of our biographical subject, it seems natural that it was taken directly from the Greek elektron. In reality, though, the word was first used in the 1850s as an alternative for electrum – a direct parallel with the Latin – which refers to an alloy of gold and silver. It moved into science in 1891 when the Irish physicist George Stoney used it (after briefly dallying with electrolion) not as a reference to a physical entity, but a unit of electricity, which would gradually be regarded as an object in its own right. An electrically charged atom has been referred to as an ion since the 1830s, so a separate charge was an electro-ion – an electron.

A mysterious attraction

How amber was able to attract things was just as much a mystery to early natural philosophers as a magnet's invisible influence. An attractive force was always a difficult proposition to explain because it's much easier to push something remote away (say by blowing it) than it is to pull it towards you. Isaac Newton would later have the same problem when describing the action of gravity – he was mocked by some of his contemporaries for making use of the term 'attraction', because the word was largely associated with personal feelings, not a force. It sounded as if the Moon fancied the Earth.

Newton attempted to explain gravitational pull by imagining a universal flow of particles that were constantly pushing bodies in space. If something like the Earth was placed in the way of that flow, blocking particles from its direction, then the particles pushing from everywhere else would naturally tend to make an object floating in space move towards the Earth, giving the effect of attraction.

---

* Rather oddly, 'calefy' means to make warm or hot: as we will see, this was the result of a confusion that dated back to the ancient Greeks.

When the first-century CE Greek philosopher Plutarch described the attractive ability of amber he made use of a similar argument. In his *Platonic Questions* he attempted to explain amber's abilities:

> And neither amber nor the loadstone draws any thing to it which is near, nor does any thing spontaneously approach them. But this stone emits strong exhalations, by which the adjoining air being impelled forceth that which is before it; and this being carried round in the circle, and returning into the vacuated place, forcibly draws the iron in the same direction. In amber there is a flammeous and spirituous nature, and this by rubbing on the surface is emitted by recluse passages, and does the same that the loadstone does. It also draws the lightest and driest of adjacent bodies, by reason of their tenuity and weakness; for it is not so strong nor so endued with weight and strength as to force much air and to act with violence and to have power over great bodies, as the magnet has.*

Because heated air moves, there seemed to be an association with rubbing amber and generating heat (that 'flammeous and spirituous nature') to force air away, causing the vacuum that generated the attraction. Note the similarity to Thomas Browne's use of 'calefy'. With no concept of electrical charge or electrons (other than ambers), the ancient Greeks and their successors all the way through to Victorian times had little alternative but to identify electricity with the more understandable forces generated by heat.

Transferring electrons

We now know that amber is a fossilised tree resin, typically between 100 and 300 million years old. When it is rubbed, typically with a cloth or fur, just like the balloon in the opening chapter, it undergoes

* Plutarch's *Moralia (Platonic Questions 7.7)*. Translated from the Greek by several hands. Corrected and revised by William W. Goodwin, PH. D. Boston. Little, Brown, and Company. Cambridge. Press Of John Wilson and son. 1874.

a triboelectric process – electrons are effectively rubbed off the cloth leaving a relative negative charge on the amber, able to induce a positive charge in nearby objects, which are then attracted to it. It all seems so straightforward now ... if it weren't the case that triboelectricity and its close relation 'contact electrification', which involves producing charge without rubbing, are still not fully understood.

The gold standard for scientific studies is replication: when a study is repeated, the results should be the same. It was failures of replication that put many earlier social sciences papers in doubt from around 2012, but such failures are relatively uncommon in physical experiments, where far higher standards of significance are deployed. Even so, replication has proved elusive when dealing with triboelectricity. It's not even clear whether and why a particular substance gets a negative or positive charge when touched or rubbed. Scott Waitukaitis, a physicist working in the field, was quoted in 2025 as saying, 'We can't even get the experiments to just behave.'*

Waitukaitis and his co-worker Juan-Carlos Sobarzo ran a series of experiments hoping to see if the presence of water molecules was influencing the variability in contact electrification. In the grand tradition of static electricity experiments they failed – but along the way were able to uncover a reason behind the strange behaviour. They were using small pieces of the polymer polydimethylsiloxane, some of which, after contact with other small pieces of it, developed a negative charge.

True to form, the way this happened was not consistent, yet with repeated contact it became so. It appeared that repeated contact had an effect on the way that a square of polymer became charged. The more contacts there were, the more likely that a sample would get a negative charge. Exactly how this was possible – that somehow the bits of polymer developed a memory – was not clear until the team examined their samples with an atomic force microscope and found that the samples that gained a negative charge had smoother surfaces, with irregularities worn down by the repeated contact.

---

* Scott Waitukaitis was quoted in *Solving the 250-year-old puzzle of how static electricity works* by Elise Cutts in *New Scientist*, 25 July 2025.

At the time of writing, it's still not clear why this is the case. But the discovery that repeated contact changes the ability of a material to gain a charge helps explain why there has been so much inconsistency in studies on triboelectricity and contact electrification.

However amber gets its charge, though, its role from ancient times was more inspirational than scientific in the awareness of electricity, something that would still be a mysterious concept in Newton's day.

NEWTON'S SPIRIT AND THE ELECTRICAL BOY

4

How far strictly speaking the Spirit in this Operation may be said to be fired by the repulsive Power of Electricity, or how far that Power, which repels light substances when fully impregnated with Electricity, fires the Spirit, may be the Subject of a future Inquiry ...

William Watson, *Electrical experiments – letter to the Royal Society* (1745)

Isaac Newton is, after Einstein, the best-known scientist in history. Born in Woolsthorpe Manor in Lincolnshire (a grandly named large farmhouse) on Christmas Day 1642,* Newton had a miserable upbringing – his widowed mother remarried when he was three and left him with his grandparents to live with her new husband. It was only when Newton got to Cambridge University that he seemed to truly come into his own.

Like Benjamin Franklin, science only took up a small part of his life – his studies took in alchemy and theology, while he would become master of the Royal Mint and a member of parliament. Despite the often-made claim that he was the first person to be knighted for his contribution to science, his knighthood was for his management of the currency. Another parallel with Einstein is the mythology that has built up around Newton, notably that of the apple tree.

* Newton's birthday is often marked on 25 December, but in the modern calendar his birthday would be 4 January 1643.

The real apple

The doubtful version of this scientific legend is that Newton was sitting under an apple tree on the family farm when an apple landed on his head, inspiring him to think about the nature of gravity. So enthusiastic are some to dismiss this certainly exaggerated account that they dismiss any intervention of an apple at all. This goes too far the other way, as Newton himself described the way that an apple falling (not on his head) made him think.

In *Memoirs of Sir Isaac Newton's Life*, published in 1752 by antiquarian William Stukeley, we are told of a visit to Newton's lodgings in Orbol's Buildings, Kensington, on 15 April 1726. Stukelely writes: 'After dinner, the weather being warm, we went into the garden, and drank thea under the shade of some apple trees; only he and myself. Amidst other discourse, he told me, he was just in the same situation, as when formerly, the notion of gravitation came into his mind. Why should that apple always descend perpendicularly to the ground, thought he to himself; occasion'd by the fall of an apple, as he sat in a contemplative mood.'

What we know of Newton's exploration of electricity is more direct, though limited. His working life was just a little too early for the transition of electricity from a small phenomenon associated with amber and a few other materials to become a significant part of physics. Newton's two great books, *The Principia* (*Mathematical Principles of Natural Philosophy*) and *Opticks*, both include passing references to electrical matters, but these were largely relegated to secondary sections of the books.

In the General Scholium

The Principia is where we get Newton's work on forces and gravity. For its second edition, published in 1713, he added some extra material. Where the final volume had originally ended without any overall conclusion, for this version he added a 'General Scholium' (*Scholium Generale*). It is this addition that contains one of *The Principia*'s most famous lines, 'Hypotheses non fingo'. This is translated as 'I frame no hypotheses' – i.e. I'm just reporting on what happens, but I'm not hypothesising on how or why it happens. It's thought, however, that the word 'fingo' had negative connotations at the time, suggesting cobbling something together without evidence. This is more in the sense of framing someone than of producing a scientific argument.

In the General Scholium, Newton refers to 'a certain very subtle spirit pervading gross bodies and lying hidden in them'.* He tells us that 'by its force and actions, the particles of bodies attract one another at very small distances and cohere when they become contiguous; and electrical bodies act at greater distances, repelling as well as attracting neighbouring corpuscles ...' From other notes Newton made, this all appeared to be referring to an electrical phenomenon.

Newton was influenced in writing this by experiments undertaken by Francis Hauksbee – a former draper who became, in effect, Newton's pet experimenter through the Royal Society. The Society, Britain's leading scientific institution, had played a major role in Newton's scientific growth. He was never a sociable individual, but had corresponded with the Society over the years, if occasionally falling out with key figures, and would become its president.

Hauksbee had moved on from the traditional use of amber in electrical experiments, primarily making use of a glass globe from which air had been removed as much as possible with the primitive air pumps of the day (Hauksbee designed his own). This ball, containing a little mercury, was rotated in a wooden frame – when a hand touched it to rub against the moving surface, a glow developed in the globe.†

This so-called barometric light had been noticed back in 1675 by the French astronomer Jean Picard, when a glow came from a moving mercury barometer, but Hauksbee made a much more thorough investigation of the phenomenon. This was the initial basis of

---

* Extracts from Newton's General Scholium and notes taken from *The Principia – a new translation* by Bernard Cohen and Anne Whitman (1999).

† Hauksbee's glass friction device was not the first – German scientist Otto von Guericke had experimented with rubbing a ball of sulphur, though there probably wasn't rotation involved, and von Guericke was not considering electricity in the same sense but rather the fuzzy concept of 'cosmic potencies'.

Newton's description of an 'electric spirit'. He would suggest that the light emitted in the experiment (and by hot bodies) was a result of the electric spirit being 'sufficiently agitated'. In later writings, Newton considered a wider connection between electrical spirit and light, even suggesting that electric spirit that was embedded in glass was somehow responsible for the way that light was made to change direction when reflecting off mirrors and refracting through prisms and lenses.

This was the kind of speculation that Newton on the whole tried to avoid making public (one of the hypotheses he didn't want to fingo) and he appears never to have published it. Entertainingly, it is surprisingly close to the modern view. As we will discover later, both reflection and refraction are a result of light interacting with electrons in matter – and the fuzzy presence of those electrons with no fixed location can easily be considered an 'electrical spirit'. Because of his wariness, Newton ended (very sensibly) by saying that 'there is not a sufficient number of experiments to determine and demonstrate accurately the laws governing the actions of this spirit'. This would remain the case for a couple of centuries.

Where *The Principia* had been published in Latin, Newton's later title on light and optical effects, *Opticks*, was originally published in English, and here electricity featured again. He mentions 'the electrick Vapour which is excited by the friction of the Glass against the Hand ...',* referring to Hauksbee's work, and adds that its effects have also been found 'by rubbing a long and large Cylinder or Glass or Ambar [sic] with a Paper held in ones hand ...'

He went on to describe what was happening as how 'an electrick Body can by Friction emit an Exhalation so rare and subtile, and yet so potent, as by its Emission to cause no sensible Diminution of the weight of the electrick Body ...'

Dealing with Newton's ideas on electricity can feel a little like wading through the treacle of obscurity, but bear in mind how little was known about it at the time. Even though electrons had been operating throughout the universe for 13.8 billion years, humanity had only had

* Quotes from *Opticks* taken from the Second Edition (1718).

a short observation of some of their mysterious abilities, lacking the physical tools to investigate further.

Electricity as stunt show

While Newton was primarily concerned with what might lie behind electrical effects, others of his contemporaries were more enthusiastic about exploring the limits to which electricity could be exploited for dramatic effect, a phenomenon that would be brought to its extreme in the eighteenth century in two displays – the Electrical (or Flying) Boy, and the Westminster Bridge experiment.

The Electrical Boy was sometimes known as Gray's orphan after the experimenter behind it, Stephen Gray, a somewhat younger contemporary of Newton. Gray worked in the dyeing trade, but was both an amateur astronomer and electrical experimenter who is often credited with the discovery that electricity can flow. Although Newton referred to his electrical spirit as a kind of fluid, electrical experiments had largely been confined to static electricity, where an electrical charge is built up on an object, such as a piece of amber. The only movement involved was either of attracted small objects, or of a spark.

For Gray, though, the discovery of what become known as current electricity was what made the subject worthy of exploration. Gray discovered that he could extend an electrical source with twigs or oily thread and still get the same effects remotely from the original charged object. By 1729, he had also discovered the importance of insulators to prevent this mobile electricity from returning to the Earth – if a conducting thread was held up by silk, it would make a much longer transmission possible. At a friend's house in Kent, Gray was able to run a conductor for 800 feet (250 metres) and still produce an electrical effect.

This was all communicated to John Desaguliers (a French-born English natural philosopher), who had taken over from Hauksbee as the Royal Society's demonstrator. Desaguliers had been an early supporter of Gray and came up with the terms conductor and insulator. Gray also noticed that it was possible to pass the electric 'virtue', or spirit, from one object to another without direct contact.

Although not realised at the time, this electrical induction was the mechanism behind the ability of rubbed amber to attract a piece of paper.

The Electrical Boy was Gray's most dramatic demonstration.* A boy was suspended from silk ropes to insulate him. A source of electrical charge rubbed against the feet of the boy, whose outstretched hands then demonstrated electrical abilities. In one of the most dramatic versions of what can only really be described as a stage act, illustrated in a book by William Watson, the boy transmits electricity to a girl standing on a tar-covered (hence insulating) barrel, using her other hand to attract feathers or bits of paper. Initially, this was done using glass tubes, rubbed with material, but the later version, illustrated by Watson, used a rotating electrostatic generator, able to create a strong enough charge to produce sparks from the subject's fingers.

This was certainly as much entertainment as science, though even the Electrical Boy would be written up in the prestigious *Transactions of the Royal Society*, where Gray gave details in the manner of an experimental write-up, noting that he used 'a Boy between eight and nine Years of Age. His Weight, with is Cloaths on, was forty-seven Pounds ten Ounces'.

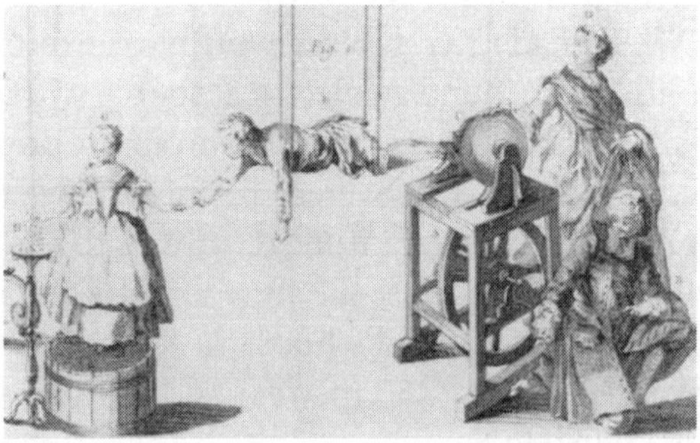

The electrical boy.

Public domain via Wikimedia Commons

* One source refers to Gray's Electrical Boy as the 'Flower Boy', as one of the possible objects attracted using electricity was flower petals. This feels somewhat disappointing compared to the Electrical or Flying Boy.

William Watson, who publicised the experiment in his 1748 book *Experiences et observations pour servir a l'explication de la nature et des proprietes de l'electricite,* was a medical doctor who was part of the next generation of electrical discovery. He took the combination of electricity and humans to a new level in the mid-1740s by connecting a source of electrical charge, via a man (presumably standing on something insulated), to a 1,200 foot (365 metre) wire which traversed the Thames across Westminster Bridge. At the end of the wire another man linked it to the waters of the Thames, while a final person had his hand in the water on the opposite bank of the river and connected it back to the source. All three felt the shock of electrical transmission.

A jar of electricity

Watson was using a next-generation source of electrons. The earlier experiments and demonstrations had relied on direct connection to a means of generation, whether rubbing a glass tube or a mechanical generator that continuously rubbed a rotating disc. But it would prove far more convenient if the electrical 'fluid' could be stored away in a jar, as if it were water. Two Europeans, the German priest and natural philosopher Ewald von Kleist and the Dutch natural philosopher Pieter van Musschenbroek, apparently independently, around 1745 came up with the idea of keeping electricity in a jar, part filled with water, with a brass rod pushed through the lid into the water.

That same year, Watson introduced this concept to the Royal Society in London, and it was such a jar that he used as the power source for his Westminster demonstration. Initially such jars were often referred to as Kleistian jars, but they were publicised (and probably sold) by the French priest and scientist Jean-Antoine Nollet, who called them Leyden jars after van Musschenbroek's academic home, Leiden University. Nollet, incidentally, had his own demonstration to rival Watson's. As a clergyman he was able to arrange for 200 monks to stand in a circle in a monastery's grounds, linking them with wires to a Leyden jar in an attempt to measure the speed with which the electrical effect travelled round the circle. It appeared to be extremely quick, but no speed could be measured and there was little scientific benefit other than producing a large number of irritated monks.

Although a seemingly small development, the ability to store electricity, rather than generate it by rubbing each time it was required, was a major step forward in the emergence of electricity (and hence the electron) as a phenomenon with wider potential uses than entertainment alone. By 1767, another of the next-generation scientists, better known for his discovery of oxygen, would put together a definitive description of the 'history and present state of electricity'.

PRIESTLEY'S HISTORY AND PRESENT STATE

5

> *A cursory scanning of the more than one hundred and fifty titles in Priestley's bibliography shows contributions in language study, English grammar, philosophy of education, rhetoric, politics, history, religion, and biblical criticism, as well as the science for which he is best known.*
>
> Robert E. Schofield, *The Enlightenment of Joseph Priestley* (2004)

Joseph Priestley was almost the definition of a polymath – though some of his contemporaries would also have also considered him a danger to their very existence. He was certainly a character. Born in Yorkshire in 1733, he is remembered for his discovery of oxygen (which he called dephlogisticated air) in 1774, undertaken while working as a librarian for the Marquis of Lansdowne at Bowood House in Wiltshire.

From the brewery to the Lunar Society

Earlier in his career, Priestley had spent some time working on carbon dioxide, making use of the biproduct of the Jacques Brewery in Leeds. (He was later banned from the brewery when he ruined a batch of beer while trying to dissolve carbon dioxide in ether.) He was one of the first to bubble CO_2 through water, making the everyday liquid taste like fizzy mineral water from the Alps.

At the time, Priestley did not take this further, but in 1772 he was invited as a dinner guest to the Duke of Northumberland's home in London (presumably as entertainment for the more prestigious attendees). The Duke liked to add something novel to the dinner, and on this particular occasion gave his guests distilled seawater, to demonstrate that it was possible to make it safe to drink. Unfortunately, the water was also extremely bland and didn't go down well. Priestley piped up that he could make it more enjoyable to drink and next day the Duke's guests were treated to Priestley's soda water. Typically of a man whose theories exceeded his practical abilities, Priestley never got round to developing fizzy water commercially, leaving it to the Swiss Johann Schweppe a few years later.

Later in his career, Priestley's extreme version of religious nonconformism got him into so much trouble that his house near Birmingham was burned down by an angry mob. This was part of a wider series of riots in and around Birmingham, largely inspired by the support given to the French Revolution by religious dissenters – English protestants who split from the dominant Church of England. In 1791, Priestley fled to America where his religious (and revolutionary) views were less controversial. But his work on electricity took place during a calmer period of his life, while living in the Cheshire town of Warrington between 1761 and 1767. Here he had contact with many of the leading lights in natural science of the day, including at a gathering, as we have already heard, that Benjamin Franklin attended. This was the Lunar Society of Birmingham, whose members included naturalist Erasmus Darwin,* potter Josiah Wedgwood, and engineers Matthew Boulton and James Watt.

This talking shop, given its name as it was arranged on full Moon nights so the members could see to make their way home safely afterwards, played a central role in the British Enlightenment, though Priestley would indirectly be responsible for its demise, as the riots in 1791 tainted the reputations of other members with the mob, leaving some of the Lunar Society fearing for their safety. Priestley would not make any significant contributions to electricity himself, but his book *The History and Present State of Electricity*, published in 1767, provides an

* Erasmus Darwin was Charles Darwin's grandfather.

ideal (if extremely tedious) snapshot of the state of electrical knowledge in the second half of the eighteenth century, a pivotal period in the electron's prehistory.

A field full of pleasing objects

To the modern eye, Priestley's writing style is painfully prolix. In the 21-page preface, for example, we learn in one paragraph that 'The history of electricity is a field full of pleasing objects, according to all the genuine and universal principles of taste, deduced from a knowledge of human nature. Scenes like these, in which we see a gradual rise and progress in things, always exhibit a pleasing spectacle to the human mind. Nature, in all her delightful walks, abounds with such views, and they are in a more especial manner connected with every thing that relates to human life and happiness; things, in their own nature, the most interesting to us. Hence it is, that the power of association has annexed crouds [sic] of pleasing sensations to the contemplation of every object, in which this property is apparent.' Factual content practically zero.

After several pages of such meandering musing, Priestley notes, 'What inexpressible pleasure would it give a modern electrician, were the thing possible, to entertain such a man as Sir Isaac [Newton] for a few hours with his principal experiments!' This is revealing – at this stage in our understanding of electricity it was seen as much a source of entertainment as information. While knowledge of electricity's effects was growing, there was no real idea of practical use for it, or explanation of what was happening.

The 1769 edition of Priestley's book, quoted here, includes an 'advertisement' section 'Concerning the second Edition; and some Additions to it'. One short extract on new additional experiments gives a very effective feel for the way that scientific findings were described then. We are told, 'Professor Winthrop observed his apparatus to be strongly electrified for several hours, while the snow, which fell the day before (and which had not electrified his apparatus while it was falling) was driven about by a high wind. This he had observed twice before.' While we wouldn't expect a paper on an experiment at CERN today to include details of the weather, in a sense we can admire the care with which these early scientists, unsure of exactly what was involved with electricity, made sure their readers knew the prevailing conditions.

Priestley divides up electrical history into ten periods. These begin with experiments and discoveries 'prior to those of Mr Hawkesbee', before going into individual details of the work of a number of characters we have already met, from Gray and Desaguliers to Watson and Franklin. (I am using Priestley's spelling of names here, such as Hawkesbee rather than Hauksbee, which sometimes differs from modern renditions.) He then moves on to the theories of the day, the construction of electrical machines, practical maxims for the use of young electricians and a description of the most entertaining experiments performed using electricity.

A lot of Priestley's book consists of detailed descriptions of experiments that were undertaken and what was observed. Take Gray's experiments at his friend's house in Kent mentioned in the previous chapter. An extract gives a vivid feel for the mix of detail and vagueness in the description of the experimental process:

> The first experiment ... was in a matted gallery at Mr. Wheeler's house – July 2d 1729, about ten o'clock in the morning, as Mr. Grey, after his usual manner, has minutely recorded it. About four feet from the end of the gallery, they fastened a line across the place. The middle part of this line was silk, the rest pack-thread. They then laid the line to which the ivory ball was hung, by which the electric virtue was to be conveyed to it from the tube, and which was eighty feet and a half in length, across this silk line, so that the ball hung about nine feet below it. The other end of the line was, by a loop, fastened to the tube, which they excited at the other end of the room. After this preparation, they put the leaf brass under the ivory ball, and, upon rubbing the tube, it was attracted, and kept suspended for some time. The gallery not permitting them to go to any greater lengths with a single line of communication, they contrived to return the line, making the whole length of it almost twice that of the gallery, or about one hundred and forty seven feet, which answered very well. But, suspecting that the attraction would be stronger without doubling or returning the line, they made use of a line one hundred and twenty four feet long, running in one direction in the barn; and, as

> expected, they found the attraction stronger than when the line had been returned in the gallery.

After several hundred pages of this kind of description we can begin to uncover the theory that was most commonly held at the time. Priestley points out the care that is needed in developing hypotheses, particularly in a field like this: 'Electricians have generally been too much attached to their several theories, so as to have retarded the progress of real discoveries. Indeed, no other part of the whole compass of philosophy affords so fine a scene for ingenious speculation … As the agent is Invisible, every philosopher is at liberty to make it whatever he pleases, and ascribe to it such properties and powers are most convenient for his purpose.'

From experiment to theory

Priestley takes us through earlier theories before telling us that he will content himself 'with giving the outlines of some principal theories of electricity which have adherents at present …' Most of these depended on an electrical fluid, with the relatively modern addition at the time of dividing electricity into positive and negative kinds. Particles of the fluid repelled each other, but attracted 'electric' substances (such as amber and glass). The fluid was thought to be able to move through conductors, because they had larger pores than insulating substances like glass.

The meaning of positive and negative in the electrical context, primarily the work of Benjamin Franklin, is described; for example, in an experiment where one person rubs a glass tube and another touches it to receive a shock: 'The person who touched the tube as said, by Dr Franklin, to have electrified positively, or plus; being supposed to receive an additional quantity of electrical fire: whereas the person who rubbed the tube was said to be electrified negatively, or minus; being supposed to have lost a part of his natural quantity of the electric fluid.'

Oddly, as described here, the 'electrical fluid' appeared to move from negative to positive, where the convention in electricity would become a flow from positive to negative because a positive location was supposed to have excess fluid, which would then flow to the negative. (In reality, electrons flow in the opposite direction from negative to positive, more like Franklin's 'electrical fluid'.)

We should bear in mind the parallels here with the way that heat was understood at the time. Heat was also considered to be an invisible fluid, called caloric, that moved from a hot body to a cooler one when they were in contact. This was a good example of an incorrect theory that worked well to describe what was observed, until later experiments, which seemed to show that caloric never ran out, demonstrated that it could not logically exist. The electrical fluid was a natural extension of the concept and equally worked well initially as an explanation.

Our natural inclination on hearing that Priestley was giving information on the construction of electrical machines might be to assume that he was describing some sort of electric motor – but that was some way off in the future. The electrical machines of Priestley's day were for the generation and storage of electrical charge (some of which looked more like spinning wheels than anything else). Having said that, some of the 'most entertaining experiments performed by electricity' do involve producing motion by making a glass ball rotate, or producing an air current from an electrical discharge to turn vanes. Not practically usable as motors, but a first indicator that electricity had the potential to act as motive power.

For now, though, the direct, personal, physical experience of the shocking nature of electricity would continue to be at its heart. As Priestley tells us: 'A great deal of diversion is often occasioned by giving a person a shock when he does not expect it; which may be done by concealing a wire that comes from the outside of the phial* under the carpet, and placing the wire that comes from the inside in such a manner in a person's way; that he can suspect no harm from putting his hand on it, at the same time that his feet are upon the other wire … When a single person receives the shock, the company is diverted at his sole experience; but all contribute their share to the entertainment, and all partake of it alike, when the whole company forms a circuit by joining hands …'

* When Priestley refers to a 'phial' he is talking about a Leyden jar.

Overall, Priestley gives us a picture of a mysterious and entertaining phenomenon for which some initial theories are beginning to be put together, with many experiments used to show how different materials may be used for generation and conduction of electricity, including an increasing experience of being able to store some charge in a precursor of an electrical battery.

Electricity was also starting to be recognised in a natural context, but since Newton there had been relatively little speculation about the relationship between electricity and life. Until an Italian natural philosopher brought together the unlikely combination of an electrical circuit and frogs' legs.

THE AMAZING TWITCHING FROG

6

When I was about fifteen years old, we had retired to our house near Belrive, when we witnessed a most violent and terrible thunder-storm. It advanced from behind the mountains of Jura; and the thunder burst at once with frightful loudness from various quarters of the heavens. I remained, while the storm lasted, watching its progress with curiosity and delight. As I stood at the door, on a sudden I beheld a stream of fire issue from an old and beautiful oak, which stood about twenty yards from our house; and so soon as the dazzling light vanished, the oak had disappeared, and nothing remained but a blasted stump … The catastrophe of this tree excited my extreme astonishment; and I eagerly inquired of my father the nature and origin of thunder and lightning. He replied, 'Electricity;' describing at the same time the various effects of that power. He constructed a small electrical machine, and exhibited a few experiments; he made also a kite, with a wire and string, which drew down that fluid from the clouds.

Mary Shelley, *Frankenstein, or, The Modern Prometheus* (1818)

When Newton referred to an electrical 'spirit' it is hard not to see some kind of connection with the idea of electricity as a kind of 'vital force' that could be connected with life itself – but it was not really until the eighteenth and nineteenth centuries that such a connection would be made with a more detailed scientific view. Immersed in modern culture it's easy to tie this into Mary Shelley's remarkable novel *Frankenstein*, but as we have already seen, the electrical aspect of bringing the creature to life originates with the 1931 film, not the original story.

The quote at the opening of this chapter is the only reference to electricity in the whole of the book. This may be something that Mary picked up from her husband-to-be, as the young Percy Shelley had enjoyed 'electrifying' his family for amusement. Percy's sister Hellen would describe this as involving significantly less enjoyment (at least on her part) than Priestley had promised: 'When my brother commenced his studies in chemistry and practised electricity upon us, I confess my pleasure in it was entirely negatived by terror in its effects. Whenever he came to me with his piece of folded brown packing-paper under his arm, and a bit of a wire and a bottle (if I remember right), my heart would sink with fear at his approach; but shame kept me silent, and, with as many others as he could collect, we were placed hand-in-hand around the nursery table to be electrified.'*

A one-hit wonder?

Central to the development of the idea of a link between electricity and a life force would be the 1780 experiments undertaken by Luigi Galvani. According to the modern University of Bologna, Galvani was 'the most illustrious Bolognese scientist of the eighteenth century'. (Cynics might say that most of us would be hard pressed to identify another Bolognese scientist of the period.) It's perhaps a little unfair to think of Galvani as a one-hit wonder, as he had a significant impact on Italian medicine, but the field then was anything but scientific, and for most of us, if we have heard of him at all, the image that comes to mind is of twitching frogs' legs, powered by electricity.

Born in 1737, Galvani rose to become a professor of anatomy in 1772 and then obstetrics in 1782. While still primarily focused on anatomy, he became increasingly interested in work being carried out on medical applications of electricity by two Italian scientists, Leopoldo Caldani and Giuseppe Veratti. Caldani probably had most

* From *The Life of Percy Bysshe Shelley*, ed. Edward Dowden (1906).

impact in his day with his widely used medical textbooks, though he was arguably a pioneer in experimenting on spinal cords. Veratti is better known as the husband of Laura Bassi, both the first female scientific university lecturer and the first female member of a major scientific body: her interest was primarily Newtonian physics, but with Veratti's anatomy background, their joint work would also help inspire Galvani.

De viribus electricitatis

Galvani's masterpiece, which won him international fame, was *De viribus electricitatis in motu muscolari. Commentarius* (On the forces of electricity in muscular motion. Commentary), presented to the Accademia delle Scienze in 1789 and published in 1791, seven years before his death. One modern commentator has described this as having 'an impact on the scientific audience comparable to the social and political impact of the French revolution in those same years'.* This feels an exaggeration, but it did without doubt cause a stir.

It's worth briefly looking back at Newton's speculation on the electric spirit in the queries attached to his book *Opticks*. In query 24 he ponders, 'Is not Animal Motion perform'd by the Vibrations of this Medium, excited in the Brain by the power of the Will, and propagated from thence through the solid, pellucid and uniform Capillamenta of the Nerves into the Muscles, for contracting and dilating them?' The medium in question is the aetherial 'spirit' – translate the 'vibrations' into electricity and we have a very similar picture to that imagined by Galvani to explain his experimental observations.

His paper described four experiments, involving the contraction of a frog's thigh muscle when an electrical current was applied from various sources. In a neat precursor of the 1931 *Frankenstein* film, these electrical mechanisms included 'atmospheric electricity'. Inspired by Franklin's work on lightning, Galvani put one of his semi-dissected

---

* Marco Piccolino in *Animal electricity and the birth of electrophysiology: The legacy of Luigi Galvani* (1998)

Galvani's atmospheric electricity experiment.

https://commons.wikimedia.org/wiki/File:Galvani_frog_legs_experiment_setup.png

frogs on an outdoor table during a thunderstorm with a long wire from a nerve leading up to what looks like a chimney in the illustration in the *Commentarius*. He noted, 'in correspondence of four thunders, contractions not small occurred in all muscles of the limbs'.

Dramatic though this was, it would indirectly inspire the most surprising observation that Galvani was to make. To see if there was an effect from atmospheric electricity without a storm present, he repeated the experiment on a clear day with no twitching observed until he accidentally produced a result by pushing on the metal hooks that were inserted into the frog to carry the current. Further investigation showed that a movement could be engaged simply by making an electrical connection from the spine to the muscle without any external source of electricity, which, Galvani suggested, meant that animals made use of electric spirit, stored in muscle and conducted from inside to outside the muscle to elicit movement.

The effect was dramatic indeed – being able to stimulate the movement of a frog's leg with no living animal attached must have been astonishing to observers of the day – and Galvani's ideas were initially popular. As we have seen, earlier electrical work had discovered a number of non-metallic conductors such as wet string, and Galvani was also able triumphantly to reproduce his work without the use of metals. He produced an effect both with a slice of muscle and even with dampened paper as the conductor. But he came up against opposition from the other great Italian scientist of the period (Milanese in this case), Alessandro Volta, of whom we will discover more in the next chapter.

Although Volta initially supported Galvani's work, he would soon suggest that the electricity required to produce a movement without any apparent external source when the spine and muscle were connected came not from the animal's system but rather from within the metal used to make the link. He found that effects were stronger when using compound conductors where two metals were attached to each other (something we will come back to in the next chapter). In Volta's view, muscles were capable of responding to electrical impulses, but there was no internal electrical impulse involved in everyday muscle movement.

From Galvani to Aldini

Given the dramatic nature of the frog legs experiments, able to provide entertainment as much as scientific learning, it was probably inevitable that Galvani's death in 1798 would not result in an end to experiments on animal electrification that were based on the concept that electrical impulses played a role in the animation of living organisms. His nephew Giovanni Aldini continued with electro-muscular demonstrations, attempting to go further and bring animals, and even humans, back to life with electrical stimuli. Newgate Prison in London, the infamous site of public hangings and so no stranger to morbid spectacles, was the source of the grisly material for Aldini's most infamous show. According to the *Newgate Calendar*, when Aldini applied his galvanic process to the body of executed prisoner George Foster, which had been 'conveyed to a house not far distant' in 1803:

> On the first application of the process to the face, the jaws of the deceased criminal began to quiver, and the adjoining muscles were horribly contorted, and one eye was actually opened. In the subsequent part of the process the right hand was raised and clenched, and the legs and thighs were set in motion. Mr Pass, the beadle of the Surgeons' Company, who was officially present during this experiment, was so alarmed that he died of fright soon after his return home.

The 'house' in question seems to have been the anatomy theatre at the Royal College of Surgeons, more suited to this work than what sounds like a domestic dwelling.

Ironically, Aldini made use of an invention of Volta – the electrical pile – to produce his electrical stimuli on humans (more on this in the next chapter). Aldini's demonstrations are the link often made to suggest an influence on Mary Shelley: she does appear to have read something of Galvani's work, while it is just possible that she was aware of Aldini's dramatic attempts at revivification, though she was only five when the demonstration that led to Mr Pass's demise was undertaken, and there was no readily available account that she is likely to have read, nor does she make direct use of it in her storyline.

It is often said that in the preface of *Frankenstein* Mary Shelley wrote that galvanism had proven that life could be revived. There are two problems with this assertion. Firstly, Percy Shelley (who, as we have seen, was a bit of an electricity fan) wrote the preface, not Mary. And, perhaps more importantly, the preface contains no such reference. Galvanism *was* mentioned in the author's introduction, added later in 1831, where she says, 'Perhaps the corpse would be reanimated; galvanism had given a token of such things,' a much weaker link than is usually suggested.

It is likely that in the long-term, Galvani has done more for language than for science. Because of his work (and its dramatic nature), his name has been brought into wide usage, from 'galvanic skin response' to 'galvanisation' for the process of applying a metal coating using electricity, and the analogous use of 'galvanise' to mean to give someone energy – coming back to electricity as life force.

Later research would show just how big a part electrical charges and currents had to play in the workings of biological cells and the nervous system (though much biological electricity is due to the flow of ions rather than electrons), but it would be a different type of cell that would become an essential for electrical work by proving a more reliable source of electricity than the old Leyden jar. It was a result of the work of Galvani's would-be nemesis, Alessandro Volta.

7

VOLTA'S PILE

In this manner I continue coupling a plate of silver with one of zinc, and always in the same order, that is to say, the silver below and the zinc above it, or vice versa, according as I have begun, and interpose between each of those couples a moistened disk. I continue to form, of several of this stories [sic], a column as high as possible without any danger of its falling.

Letter from Alessandro Volta to Joseph Banks,
President of the Royal Society, 1800

If Alessandro Volta had proven something of a thorn in Galvani's side, he was to make a more significant contribution to the understanding of electricity by providing a key tool for undertaking electrical experiments. Where Galvani's name was adapted into common usage in terms like 'galvanise', Volta's name was given the far greater scientific honour of being the basis of one of the standard scientific units – the volt, which is the measure of 'potential difference' (more on what this is soon).

Volta was born in Como, then in the Duchy of Milan, in 1745. From a noble family, he would become a professor of physics, first in Como and then Pavia. Like most of the scientists of the period who made a contribution to our understanding of electricity and its use, the work of the electron was not his sole study. In 1778, Volta would isolate a substance from marsh gas, making him the discoverer of methane (the name would not be coined until the mid-eighteenth century). But electricity became his central interest.

Batteries and piles

At the time of his writing (in French) the letter to Joseph Banks quoted at the start of this chapter, Volta had already been at the University of Pavia for over 20 years. He describes his 'new apparatus' as '30, 40, 60 or more pieces of copper, or better of silver, each applied to a piece of tin, or still better of zinc, and an equal number of layers of water, or of some other liquid which may be a better conductor that simple water, such as salt water, lye, &c., or of bits of card or leather, &c, soaked in such liquids'.

Slightly oddly to the modern eye, Volta describes this setup as one that 'imitates so well the effects of the Leyden phial or electrical batteries'. Volta's pile of discs was a battery in the modern sense, but the 'battery' he referred to here was an approach taken by Benjamin Franklin of connecting together a number of Leyden jars to increase the amount of charge that could be stored: Franklin had called this a battery, reflecting the way that a battery of cannons described a set of artillery pieces collected and fired together to produce a more powerful barrage.

Volta (who writes in the third person) notes that, 'This apparatus, as it resembles more the natural electric organ of the torpedo or of the electric eel than the Leyden phial and the ordinary electric batteries, M. Volta calls the artificial electric organ.' The torpedo in question was not an underwater missile fired from a submarine, but rather a species of ray found in European waters capable of delivering an electric shock of around 200 volts. The word 'torpedo' comes from *torpere*, the Latin verb 'to numb'.

The inspiration for drawing this parallel with a fish came from a paper by English chemist William Nicholson that gave details of the structure of the fish's electricity-generating organ. This, he said, was composed of hundreds of thin columns, which might suggest a way to construct an electrical machine. Galvani had also written a memoir based on experiments with electrical fish. Now, Volta was working on the way that having two different metals in contact could act as a 'motor' for electricity. Inspired also by another idea of Nicholson's, describing an electrical 'doubler' comprised of metallic discs which was intended to intensify electrical flow, Volta's electrical pile transformed electrical experiments from around 1796.

Initially, the assumption had been that galvanic electrical effects were due to some imbalance in electrical fluid being resolved. But Volta began to conceive of something closer to a modern electrical circuit with its flow of electrons, believing that there could be a continuous current flowing as long as there was a complete circuit that did not require some external 'push' to keep it going. This, Volta thought, happened with his electric device due to an intrinsic tendency of some metals to force the electrical fluid into damp conductors. He was dealing with what we would now think of as current electricity as opposed to the static electricity (which Volta referred to as stagnating) that had been the main subject of electrical study up to this point.

Volta had moved away from Galvani's strips of metal to more practical coin-like discs, realising (unlike Nicholson) that it was paired metals in the device itself that were producing electricity. He also correctly dismissed Nicholson's idea that to build up more capacity the pairs should be linked in parallel with each other, which is the way that Leyden jar batteries were arranged. This seemed a more stagnating arrangement – for Volta's concept of circuits with flowing electricity, it made more sense that the metallic pairs were linked in series. Initially, this was in a structure known as a crown of cups, effectively a series of vessels containing liquid conductors linked by bi-metal strips, modelled on one of Galvani's experiments. Volta found this better for demonstrations, but the more important version of his invention stacked pairs of discs on top of each other separated by the liquid conductors. This was the electrical pile – a term that would become common from the French.

From contact potential to EMF

Exactly how the pile worked was not settled. Volta put it down to 'contact potential'. The idea was that having, say, zinc and silver discs in direct contact produced a tension, rather like that in a spring, which was able to force the electrical fluid through the adjacent wet conductor. It would not be long, though, before the prevailing theory became chemical, where the zinc discs interacted chemically with the wet conductor, producing electrical charge that was conducted onwards by the silver – something closer to our present understanding of the electrochemical nature of batteries.

Although the contact potential would not continue as a description of the mechanism, it is arguably the basis of two standard concepts in the physics of electricity – electromotive force (or EMF) and potential difference. These are both analogies from mechanics. Potential difference is based on the concept of potential energy. If we think of an object dropped from a height, the higher it is, the more potential energy it has – the further it drops, the more energy it has available to translate into other forms such as kinetic energy, heat, noise and impact.* Similarly, the potential difference (voltage) can be imagined as the amount of difference in potential between the two terminals of a battery (or anything else that will deliver electricity).

Volta's contact potential was based on an analogy with a compressible fluid – that mysterious stuff that was electricity (now we know effectively a bunch of electrons). The most familiar compressible fluid we know is air. Squash air up in a cylinder and it pushes back more and more – Volta imagined the electric fluid being compressed to provide the tension produced by the metals in contact, again, in effect the potential difference.

The EMF concept continues the fluid analogy, being the electrical equivalent of the work done on water by a pump as it moves the fluid. This makes EMF the energy transferred (in physics, 'work' is the transfer of energy) into a circuit per unit of charge transferred (with our electron viewpoint, this is per collection of electrons). In his letter to Banks, Volta referred to the *'force mortice électrique'* as the force to propel electrical fluid by the contact potential. Once again, this is now

---

* Perhaps the best-known myth about potential energy is that there is so much provided by, say, the height of the Empire State Building that a coin dropped off the top would kill you. Assuming this is from the observation deck, a coin would reach its terminal velocity of around 28 metres (92 feet) per second – not as fast as you might imagine, due to air resistance. That's around 100 kph (or 60 mph) – painful but not fatal. In practice, an experiment dropping coins from a weather balloon showed they fell even slower, as they fluttered around rather than falling smoothly. Physics professor Louis Bloomfield described the impact as like heavy raindrops.

measured in volts, where one way to look at the volt is joules (energy) per coulomb (electrical charge).

An electrochemical process

We now know that Volta's pile worked because of the electrical nature of chemistry. An electrical reaction between the brine (or, better, dilute sulfuric acid) in a conducting layer and zinc releases electrons that head for the copper disc where the copper catalyses their reaction to produce hydrogen. With a completed circuit, the net result is a flow of electrons from the negative zinc to the positive copper, running through a wire where it can be used to produce electrical effects.

Though wet cells with electrolytes remain with us in the likes of the lead-acid car battery, by far the most familiar are likely to be so-called dry cells like the zinc-carbon battery and the alkaline battery. While called dry, these still use an acidic electrolyte, but in the form of a paste rather than a liquid. In the zinc-carbon form, the outer casing of the battery is zinc forming the negative electrode, with a positive carbon central electrode. The active electrochemical parts are the zinc and a manganese dioxide layer that sits between the carbon and an ammonium chloride electrolyte.

Although zinc-carbon batteries still exist, the most common disposable battery is now the alkaline variety, which has both a longer shelf life and stores more energy. The only significant difference is that instead of an acidic ammonium chloride electrolyte, these use the alkaline potassium hydroxide. As a result, although the central terminal remains positive for compatibility, the zinc is now a separate central column, while the cap that sits on the carbon electrode in a zinc-carbon battery is now attached to the outer casing.

Though disposable batteries remain widely used, it is rechargeable batteries that have been central to an electrical revolution – whether it's powering a smartphone, a laptop or an electric vehicle. There are many types, but at the time of writing the most common for all these approaches is lithium-ion, which has enabled electric cars to become a mainstream option, compared with the clumsy, bulky rechargeable lead-acid batteries that were used in early electric vehicles.

Many lithium-ion batteries have a carbon negative electrode and a positive electrode based on a lithium-metal oxide (such as lithium cobalt oxide or lithium iron phosphate), with the electrolyte a salt of lithium such as lithium hexafluorophosphate dissolved in an organic solvent. Some more modern designs use a solid ceramic electrolyte, or a polymer such as polyethylene glycol. The resultant battery can be charged and discharged many times in a reversible cycle.

The main downside of these batteries is their potential to burst into flames or explode. The electrolytes are mostly flammable liquids, while charging cycles that are too quick (or sometimes at too low a temperature) can result in lithium buildup on the cathode that can penetrate sufficiently far to cause a short circuit and hence overheating. A huge amount of effort is being put into battery technology to produce power stores that can be charged faster, are more compact and are safer.

This is particularly important as batteries are incorporated as a fuel source for electric vehicles. This involves having a huge array of cells to store the amount of energy required. As a result, simply getting the battery charged can take a considerable time when compared with a two-minute fill-up of petrol or diesel. Though there are now high-speed chargers, there are major obstacles in getting the charge time down safely, requiring new battery technologies to be considered.

We shouldn't underestimate the importance of the development of batteries – our move away from fossil fuels is dependent on them. Electrons may be at the heart of much of our modern technology, but without a way of easily storing and transporting them, we would be tied to the leash of an electrical cable.

Even so, there were plenty of potential ways to use electricity before the development of modern batteries. In Volta's day, electricity remained more of an experimental art – but the work of a humble former bookbinder's assistant would change all of that.

SOON BE ABLE TO TAX IT 8

> *After the discovery of magneto-electricity his fame was so noised abroad that the commercial world would hardly have considered any remuneration too high for the aid of abilities like his. Taking the duration of his life into account, this son of a blacksmith, and apprentice to a bookbinder, had to decide between a fortune of £150,000* on the one side, and his undowered science on the other. He chose the latter, and died a poor man. But his was the glory of holding aloft among the nations the scientific name of England for a period of forty years.*
>
> John Tyndall in the Introduction to Faraday's *Experimental Researches in Electricity* (Introduction undated but written between 1867 and 1893)

If batteries provided one necessity for the increasing involvement of electricity in everyday life, generators and electric motors would drive the first true applications. It was Michael Faraday's pioneering work in the field that made this possible.

* Tyndall noted that Michael Faraday could have earned around £150,000 if he had profited from his work on electricity, generators and electric motors. To put that into a 2025 context, that would be around £19.5 million ($26.5 million) in terms of inflation, £117 million ($159 million) as earnings or £395 million ($536 million) as a construction project or similar.

This son of a blacksmith

Although Faraday was born in London in 1791, his family had roots in the north-west of Yorkshire. The family home, Clapham Wood Hall, still looks relatively substantial today, but the family had fallen on hard times by the time it was sold in 1800 by Faraday's grandfather Robert – in the deed of sale, Faraday's father James is listed as 'of the City of London, blacksmith' – a move that had been made out of desperation rather than enthusiasm. Michael was born in 1791 in the not exactly salubrious-sounding Newington Butts (now part of Southwark). Before Michael's birth, James had first set up trade in Outhgill, Westmorland, as village blacksmith, but had not been able to make ends meet.

The move south had led to financial stability, but cash remained limited – Faraday began an apprenticeship as a bookbinder aged fourteen. However, he was a determined self-educator, both reading books in the eclectic library that was the book bindery and attending the City Philosophical Society, where he made copious notes that he bound himself as part of his training. His boss, George Riebau, was so impressed that he showed the notes to a friend with an interest in the natural sciences, who gave Faraday tickets to attend the Royal Institution lectures of the scientific celebrity Humphry Davy.

This taste of the scientific life was made even sweeter by Faraday being recommended to Davy as a secretary when Davy's assistant was temporarily blinded – but the role did not last long. It was enough, though, to give Faraday a sense of purpose, and he began a series of applications to laboratories, culminating in an appointment as Davy's lab assistant after the previous occupant of the post, William Payne, lost his job for brawling.

This would begin a lifetime working at the prestigious London institution, initially undertaking vast numbers of painstaking experiments and then, as his role expanded, taking on lecturing. It was in 1821, aged 30, that Faraday had his first significant encounter with the electron, one that resulted in a painful outcome. Davy asked Faraday to write a summary of the current understanding of the relationship between electricity and magnetism. Rather than simply summarise the available papers on the subject, Faraday set out to recreate the described experiments. As part of this, he observed something he had not seen written up. When he sent an electrical current through a loose wire alongside a fixed magnet, the wire began to move in a circular fashion around the magnet.

He proudly wrote this up as his own finding, only to be accused of plagiarism by a leading London scientist and friend of Davy, William Wollaston. Wollaston had a theory that electricity spiralled down wires, something unsupported by experiment. Although there was no connection between Faraday's discovery and this theory, Wollaston believed that Faraday had stolen his idea. And Davy sided with Wollaston. It's hard not to see this as a decision based on social class. Although Davy's own origins were humble, he had become a society figure and played up to his image. Faraday had no airs and graces. Wollaston was, to Davy, 'one of us' in a way that Faraday never could be.

Despite the opposition from Davy, it was clear that this was an original discovery and Faraday did not look back. Two years later, the one-time bookbinder's apprentice became a Fellow of the Royal Society, British science's greatest honour. Only one Fellow voted against this: Humphry Davy. The fallout from the spat with Davy was something that seems to have discouraged Faraday from further electrical research for around ten years, during which he primarily focused on chemistry and the administrative duties that went with a new role as Director of the Laboratory. It was in this capacity that he introduced two lecture series that still continue today – the formal Friday Evening Discourses and the Royal Institution Christmas Lectures, primarily aimed at children.

Investigating induction

A decade after his run-in with Davy, Faraday was ready to return to electricity. He had heard reports that an electrical current in one wire could cause a current to flow in a nearby wire without any direct contact between the two – what would become known as electrical induction. Initially, this did not behave as he expected. It seemed reasonable that a flow in one wire might somehow leak to produce a flow in the other, so it might be expected that the induced current ran as long as the original did. But in reality, the induced current only flowed briefly when he turned the initial electrical supply on or off.

Faraday realised that the first wire was acting as an electromagnet when current flowed through it. It was the changing magnetic field produced by this as it was turned on and off that produced the induction. He found he was also able to generate electricity by moving a permanent magnet back and forth through a coil of wire. After his first discovery of electricity making a wire rotate, which was the key to making an electrical motor, he had now come up with the basis of the electrical generator or dynamo. Between them, these were the essentials for using electrical power practically, only lacking the electric light bulb to complete the earliest practical uses of electricity.

There is a story that when Faraday demonstrated his new discovery, the then British Prime Minister, Robert Peel, asked him what use it was, to which Faraday – a man who, as we've seen from Tyndall's remarks at the opening of the chapter, never showed a great interest in making money – is supposed to have replied, 'I know not, but I wager one day your government will tax it.'

There are significant doubts about the accuracy of this story. Would Faraday really have no idea of a potentially taxable use for electrical induction? For that matter, it's sometimes said that the comment was a reply to the then Chancellor of the Exchequer, William Gladstone, rather than the Prime Minister, while the savvy dig at the world of politics seems quite distant from Faraday's nature – and there are several variants in the reported quotation, including, 'Why sire, there is the probability that you will soon be able to tax it.'

Of the two possible participants, Peel is the more likely. He was Prime Minister twice, from 1834 to 1835 and from 1841to 1846, while Gladstone did not become Chancellor until 1852. Although the dynamo was not manufactured in a modern form until 1866, early generators based on induction were in use from the 1840s, which makes it seem unlikely that such an exchange could have occurred except in Peel's first term.

Experimental Researches in Electricity

Faraday would write up his electrical experiments in his biggest single contribution to the field, 'Experimental Researches in Electricity', which was read to the Royal Society on 24 November 1831. His paper, which would later be followed by a book of the same name with details of a whole range of experiments in electro-chemistry and the processes involved in batteries, described his work on induction, giving plenty of detail on exactly what occurred. Here, for example, he describes the coils he used:

> About twenty-six feet of copper wire one twentieth of an inch in diameter were wound round a cylinder of wood as a helix, the different spires of which were prevented from touching by a thin interposed twine. This helix was covered with calico, and then a second wire applied in the same manner. In this way twelve helices were superposed, each containing an average length of wire of twenty-seven feet, and all in the same direction. The first, third, fifth, seventh, ninth, and eleventh of these helices were connected at their extremities end to end, so as to form one helix; the others were connected in a similar manner; and thus two principal helices were produced, closely interposed, having the same direction, not touching anywhere, and each containing one hundred and fifty-five feet in length of wire.

There are over a hundred experiments described in the paper, which would presumably have taken a considerable time to read aloud. This would have taken place by gas light, something his work would relatively soon make obsolete.

There were also hints in later editions of his *Researches in Electricity* book of some idea of a relationship between atoms and electricity. Faraday noted, 'Although we know nothing of what an atom is, yet we cannot resist forming some idea of a small particle, which represents it to the mind; and though we are in equal, if not greater, ignorance of electricity, so as to be unable to say whether it is a particular matter or matters, or mere motion of ordinary matter, or some third kind of power or agent, yet there is an immensity of facts which justify us in believing that the atoms of matter are in some way endowed or associated with electrical powers, to which they owe their most striking qualities and amongst them their mutual chemical affinity.'

It's fascinating to see, at this time when electricity was becoming of practical use without any real understanding of what it entailed, this great mind, who had undertaken so many experiments on the influence of electricity on chemical reactions, fumbling in intellectual darkness towards the dual role of the electron as both carrier of electricity and an essential part of the atom and driver of its chemical qualities.

With the groundwork laid by Faraday it was not long before electricity was providing lighting and the electric motor would prove a clean, quiet alternative to the polluting alternatives of the day. But one of the first practical applications of electricity had a much more modern feeling – providing a communication network for the world.

THE VICTORIAN INTERNET

9

*Let me remind you that this great country was virtually created one hundred years ago by two inventions. Without them, the United States was impossible, with them, it was inevitable. Those inventions were, of course, the railroad and the electric telegraph.**

Arthur C. Clarke in *Tomorrow's Worlds – Satellites and Saris* (1971)

Take a moment to step back and think of the role the electron plays in your life. What would you list as the top five ways you use electrons? Here's my suggestion:

- In the atoms that make me up
- Making my body work
- In the power supply to my house
- In my electronic devices
- In my communications

Electronic devices and communications might seem very recent additions to that list, but there is no doubt about their importance in

* As an aside on the limits of AI, Google's generative AI, when asked about the meaning of the Clarke quote at the opening of this chapter, commented 'The statement is likely referring to the automobile and the telephone,' impressively managing to produce a failure on both counts.

our life. I've worked at home for 30 years, and for all of that time my business communications have primarily been made on the back of electrons, whether they are involved in email, the internet, wired phones or mobiles. In the early years, they even had a role in faxing. And this communication aspect is nothing new.

The electron's initial triumph

As we have seen from the previous chapter, it would be easy to think that the first big application of the electron's abilities would be in something less modern-sounding than telecommunications – perhaps electric lighting or motors based on Faraday's work. But the reality is that the Victorian equivalent of the internet would be the electron's initial triumph among human inventions. Communicating anything significant at a distance (the literal meaning of telecommunications) had always been limited by the speed a human being could travel to carry the message – for centuries on horseback on land, or by sailboat across water.

The earliest attempts to do better than this depended on the two naturally available carriers of information – sound and light. Sound travels about 25 times faster than a galloping horse, and light over 22 million times faster. But each medium is bound both by the complexity of information it can carry over a distance and the limitations of our means of detecting it. Church bells could be used to send an audible alert, while flags, fire and smoke could carry visual signals (given a clear line of sight). This made it possible to communicate a simple warning over a few miles – but that was about it.

It's likely that once Leyden jars were common there were some attempts to use the ability to send an electric current down a wire demonstrated by the likes of Gray and Watson (see chapter 4) in the early eighteenth century to provide remote communication. For example, in 1753 one C.M. wrote a long letter to *The Scots Magazine* titled, 'An expeditious method of conveying intelligence by means of electricity'. He suggested an electric machine be connected to an 'electric gun barrel' with 26 wires suspended separately above this. At the other end of each wire was a metal ball with a small piece of paper showing a letter lying below each ball. The sender was supposed to tap the wires in turn using a piece of glass or similar, making the bits of paper jump up in turn when attracted to their ball, spelling out a message.

This C.M. was later identified as a Renfrew-based Scotsman Charles Morison (or possibly Morrison) who was either a surgeon or a tobacco dealer. This vagueness is as nothing compared with the number of variants that have been told of this story. The version above seems to come directly from the magazine, but elsewhere you can find suggestions that the source of the electricity was 26 individual Leyden jars, the other end featured wires dipped into water to make bubbles, and that we don't know the name of the man involved. The 26-jar version would be painfully slow, but Morison's original sounded almost practical.

The marmite telegraph

In the 1790s, the French inventor Claude Chappe did not consider electricity, but rather tried to make the original vehicles of long-distance communication, sound and light, more effective. It perhaps helped that his uncle was an astronomer, making him familiar with the telescopes that would form part of the Chappe system of telegraphs. The idea of remote communication came to him before adopting his final mechanism, experimenting with using sound signals combined with a clock. The message sender would bang a large metal cooking pot (known as a marmite in France). Both sender and receiver had clocks with letters pasted on their faces.

To communicate, the sender banged his marmite when the second hand was on the first letter of his message. The receiver noted down the letter the second hand was pointing to on his clock when he heard it. Given the relatively slow speed of sound (it takes around three seconds to cross a kilometre or five seconds to travel a mile), the clocks would first be calibrated with a known message to allow for the sending delay.

The good thing about using sound is that it is not limited by visibility – but because sound waves are influenced by wind direction and dissipate quickly, their reliable range is less than a kilometre (perhaps half a mile), and the approach would be downright irritating for lengthy message sending. By contrast, while light did need line-of-sight, with a telescope it could work over a range of about 30 kilometres (20 miles), provided the mechanism being used to send the message was sufficiently large and visible – and it was a nice day.

In principle, this could be achieved by writing something on a large scale – but the method would be extremely ponderous. Instead, Chappe tried using a large wooden panel, dark on one side and painted white on the other. Each flash of the white panel was used as a silent equivalent of the marmite sound, again with calibrated clocks. This worked over a much greater distance (in daylight), but was still frustratingly slow to send a signal of any complexity.

One way of speeding up the process would have been to deploy several panels to provide a binary representation of a letter (and this was used in some later telegraphs). This is the equivalent of the ASCII code used by computers, where a number that is transmitted as zeroes and ones corresponds to each letter. For instance, the binary code 01000001 represents the capital letter A. For ASCII, eight binary numbers are used to provide a potential 256 characters – it would be possible to represent this with eight panels, though in practice a basic alphabet with a few special characters would require just five panels, giving 32 characters.

Chappe, however, went for a mechanically more complex method which he called a semaphore (a word derived from the Greek for sign-carrier).* He built towers with a wooden structure on top carrying two moveable arms, each with rotating end pieces that could be positioned at different angles. Every letter of the alphabet was represented by a particular position of the various wooden parts.

Semaphore was used to describe the approach of sending a signal this way, but Chappe devised a different word for the mechanisms themselves – a *tachygraphe* (approximately 'high speed writing' in Greek). A friend of Chappe's, Milo de Mélito, noted in his memoirs that he thought this word to be inappropriate (the distance travelled might be at high speed, but this did not strictly apply to the writing). De Mélito instead suggested that Chappe called it the more appropriate *télégraphe* (distance writing), a name Chappe adopted.

---

* We are more familiar with semaphore as a method for communicating over shorter distances using flags and it's natural to think that this came first, but flag semaphore was derived from Chappe's work, rather than the other way round.

A Chappe telegraph tower.

Public domain, via Wikimedia Commons

Having easy, speedy long-range communication provided an obvious boon to the French state. Just three years after Chappe's 1791 visual experiments, a chain of fifteen telegraph towers had been built from Paris to Lille, covering a distance of around 210 kilometres (130 miles). Messages could be passed in minutes that would have taken a long day's fast ride to cover on horseback. Such was the success of the mechanism that by the time Faraday was reaching his electrical peak in the early 1830s, over a thousand telegraph stations had been built around the world, often to link areas of potential military risk to capital cities.

Even so, this would not become a means of communication available to individuals or businesses. It was too slow, and visibility could always be limited by weather or darkness. (Illuminated telegraphs were used at night, but oil lamps were not bright enough for effective long-range communication.) The electron was ready and waiting in the wings to push the mechanical telegraph into obscurity. Just as the telegraph was taking off, experiments were underway on either side of the Atlantic that made use of new means of producing a more reliable electrical current to make Chappe's mechanical monsters redundant.

Beyond the mechanical

There had been earlier experiments on electrical telegraphy since Charles Morison's suggestion, notably that of German inventor Samuel von Sömmering who had set up a short-range demonstration in 1809 using a separate wire for each letter. And many others would make prototypes over subsequent decades without managing to set up a permanent network. A commercial system would take another 35 years to go live, in a race between two continents.

Bearing in mind this would be the first practical application of electricity, both the attempts to use electricity for telecommunications were initially considered the work of cranks. The better known of the two was Samuel Morse, based in New York, who started work on an electrical telegraph in 1832 – initially finding it near-impossible to get a message down a long enough wire to make the process worthwhile.

The Arthur C. Clarke quote at the opening of this chapter is particularly apposite, because the other invention Clarke considered made the United States possible was the railroad. Railways were initially strongly linked to the roll out of electrical telegraphs because they provided clear, relatively straight routes from A to B. Rather than needing to deal with the twisty routes and messy building around conventional roads, the railways provided fledgling telegraphy companies with well-cleared, privately managed spaces that cut their way through obstacles. It was very easy to lay a telegraph wire alongside the tracks.

Morse was not a typical inventor, even given the less professional approach to science and technology of the period. A portrait painter and sculptor rather than a scientist, he would develop an interest in photography after a visit to Paris in 1838, where he met Louis-Jacques-Mandé Daguerre, one of the pioneers of the field. His speculation on using electrical communication seems to have been inspired by the Boston-based doctor and amateur scientist Charles Jackson, who had been experimenting with electromagnets. The pair appear to have met on the voyage back from an earlier Paris trip that Morse made in 1832 to study exhibits in the Louvre.

An electrical code

The challenge that took most time was simply getting an electrical signal to reliably travel down a long wire. Working with American scientists Leonard Gail and Joseph Henry, along with engineer Alfred Vail, Morse solved the problem by using a series of relays – effectively electrically controlled switches, so the signal could be passed from section to section like a hot potato. But it was Morse himself who came up with the earliest version of his eponymous code, represented by a series of electrical impulses sent down a wire.

The two requirements were a way to encode the message and a way to make that message detectable at the end of the wire. The original designs used a stylus that dipped down onto a paper tape that was kept in motion by a clockwork mechanism. When the electrical circuit was closed by the sender, the stylus was pulled down, leaving a mark until the circuit was broken again. This made it reasonably easy to send two different kind of symbols – a short and a long mark. Although these tape-based receivers were initially used, operators found they could respond quicker to the sound the mechanical relays made, moving to audio reception.

Morse's design was initially based on a series of numbers, producing a true code. In a code, a word or phrase is represented by a number or word. So '31' could mean 'Buy three tons of wheat' ... or whatever your code book said it meant. This produced compact signals, but limited the flexibility of the communication. It was Alfred Vail who changed the design into a series of long and short pulses representing each letter. Arguably, we should know it as the Vail code.

To design this, the engineer made use of letter frequencies, intending to make commonly used letters easier and quicker to send and receive. This was a concept that printers were already familiar with. At the time, printers would set up a page using moveable type – metal blocks with the various letters on the end that were clamped together to form a page. Rather than have the same number of blocks for each letter, they would have more blocks for the most frequently used letters. Vail visited a local printworks to identify which letters he should prioritise.

In English, the frequency order approximates to ETAOINSHRDLCUMWFGYPBVKJXQZ. In Vail's code, those first five letters were represented by ·, -, · -, · ·, · ·. There was potential for confusion between O and I, where the only distinction was the length of gap between two dots – more obvious on paper than was the case audibly. When Morse code became standardised, O lost its shortened form (arguably it should have had two dots while I should have taken the two dashes allocated to M). It became the odd one out by having the longer notation familiar from SOS: - - -.

Morse's timing when developing his system was excellent. Wanting to do something inspired by the French telegraph system but better and more appropriate to the scale of the country, the US Treasury made it known that it was intending to develop a high-speed communication system. Morse put forward a proposal for this using electrical means and demonstrated his technology in Washington – but the funding was not forthcoming.

Following this failure, he tried to raise money in Europe – also without success. It was only when news arrived from England about a successful rival system being installed that Morse received $30,000 funding* from Congress to set up a trial electric telegraph line running alongside the railroad track between Washington and Baltimore, which had been constructed around nine years earlier, covering a distance of around 60 kilometres (38 miles).

Initial attempts were not encouraging. The iron-based electrical wiring was poorly insulated and had rusted through in the damp lead pipes it was buried in before the full length could be completed. Morse

* $30,000 in 1844 would be the equivalent of around $1.34 million with simple inflation, $20 million in terms of wages or around $30 million as a construction project. These different values reflect the increased cost of labour relative to goods, and the scale of the payment with respect to the economy as a whole.

changed tack, suspending wires from wooden poles instead. The first message sent on the system was 'What hath God wrought',* sent on 24 May 1844. Most of the construction was in the hands of Ezra Cornell, who added the essential component of glass insulators to the system to keep the wires from shorting out. Cornell built up a telegraph business that became part of Western Union, eventually making him sufficiently wealthy to set up the eponymous university in New York.

A royal birth and a murderer arrested

The wake-up call from across the Atlantic that got Morse his funding was the work of inventor William Cooke and the physicist Charles Wheatstone, who started on a similar project four years after Morse, but were able to launch their own system with relays along the way in 1838, which would soon play a dramatic role in apprehending a murderer on the loose, where trains proved no match for electrons in a race against time. Cooke and Wheatstone's system was less flexible than Morse's, but had the advantage of requiring no expertise to send and read messages. Instead of encoding the signal as dots and dashes, the UK system had a set of five pointers, each of which could be oriented left, centre or right.

The pointers were linked on a backboard to a set of letters, so each position indicated a specific letter. The sender would spell out their message letter by letter and the pointers on the receiver's end would sequentially indicate the same set of letters. Just like Morse, Cooke and Wheatstone found that a railway provided the ideal route for their wires, running initially from Paddington station in London to the town of West Drayton about 20 kilometres (13 miles) to the west. By 1843, cabling had been extended to the then small town of

* Morse's 'What hath God wrought' message is usually quoted out of context. It's taken from verse 23 of the 23rd chapter of Numbers, the fourth book of the Bible, from the Authorised Version: 'Surely there is no enchantment against Jacob, neither is there any divination against Israel: according to this time it shall be said of Jacob and of Israel, What hath God wrought!'

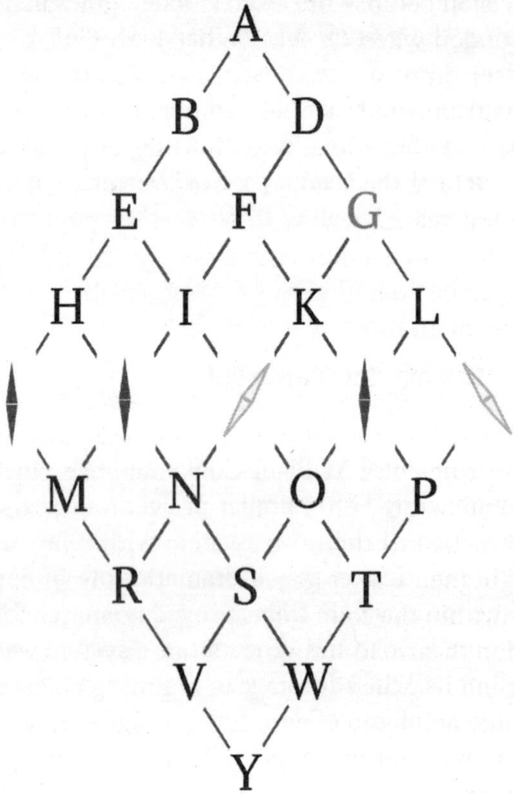

Cooke and Wheatstone telegraph, indicating the letter G.

Slough around 32 kilometres (20 miles) west of London, by now using a single indicator in a more clock-like layout.

This apparently obscure location to link to the capital was chosen because Slough was the closest railway station at the time to Windsor, one of the main homes of the British royal family, and also near to Eton with its exclusive school. (Allegedly, the railway company had wanted to build a station at Eton, but the authorities there thought their pupils would be in danger of bunking off to the fleshpots of London and vetoed it. There is still no Eton station.)

Despite the royal connection, it might still have seemed a strange route over which to start a system – but it proved an excellent marketing decision because the electric telegraph was used to send a message to London when Queen Victoria's second son, Alfred, was born at Windsor on 6 August 1844. Thanks to the high-speed connection, it was trumpeted that *The Times* newspaper had been on the streets in London within 40 minutes of the announcement being made in Windsor – rivalling the speed of a modern rolling news channel.

Less than ten years later, those high-speed electrons would provide the means to catch a murderer by the name of John Tawell. The case would become something of a cause célèbre. In a description of the subsequent trial in the edition of the *Illustrated London News* for 15 March 1845 we are told, 'The scene both inside and outside [Aylesbury] court is represented as very disorderly, from the excitement that prevailed on the subject. It more resembled the crush and rush at the hustings of a contested election than the opening of a court of Criminal Justice. After a struggle of more than half an hour's duration, destructive to clothes, and almost fatal to life or limb, the reporters fought their way to a miserable little corner of the gallery ... But the scene within the court beggars all description. The opening of a minor London theatre for the first night of a Christmas pantomime is the only scene to which it could be compared. Crushing, yelling, jeering whistling and making every sort of noise.'

London-based businessman Tawell had a decidedly chequered history. He was transported to Australia for possessing forged banknotes as a young man, where he made a success as a pharmacist and in other business ventures. Returning to London, he employed a nurse, Sarah Lawrence, to look after his wife Mary, who was suffering from tuberculosis. Mary died soon after and Tawell remarried, but he kept on Sarah (now called Hart) as his mistress and provided her with a house on the outskirts of Slough.

On 3 January 1854, after a terrible argument, he poisoned Sarah and ran to the nearby station, catching the London-bound train. Without the telegraph, a message could only have been sent to London on a later train, giving him the chance to disappear. But the police were able to alert Paddington station, asking for a man 'dressed like a kwaker' to be held on arrival. This referred to Tawell's drab clothing, likening him to a Quaker (which indeed he was), but the system's limited letter options did not include a Q. As a result of the message, Tawell was caught at Paddington, tried and executed.

Such were the obvious benefits of the electric telegraph that it soon overtook and made obsolete the mechanical equivalents. Just six years after Morse's first message was sent there were around 19,500 kilometres (12,000 miles) of telegraph cable in the US alone. The telegraph became synonymous with the transmission of timely information – so much so that when a new newspaper was launched in London in 1855 it seemed apt to name it the *Daily Telegraph*.

Crossing the Atlantic

Important though the electron's role in carrying messages would be within countries, its world-spanning potential would become clear with a dramatic development just ten or so years after Morse's first message was sent – the construction of a transatlantic cable, linking America and England, requiring around 4,600 kilometres (2,900 miles) of cable to be laid on the bed of an unforgiving ocean. This massive undertaking would bring into the field one of the greatest physicists of the age, William Thomson, Lord Kelvin.

The cable project was initially primarily financed by the New York paper magnate Cyrus Field, whose team, headed by Wildman Whitehouse (who had no experience in the technology), managed to spend a fortune on the project – around $1.4 million, on inflation alone around $55 million now and the equivalent of a construction project costing around $850 million. The copper cable was insulated using a natural substance called gutta-percha, a latex obtained from a Malaysian tree called the pertja, which would also be used to form the core of golf balls. The cable run would be from Valentia Island off Ireland to Trinity Bay in Newfoundland. In both cases these islands off the coast proved an easier grounding point than the mainland, with an easily laid secondary cable to finish the link.

In 1857, there was a disaster when the cable broke before it had been completed. The other leading Scottish physicist of the day, James Clerk Maxwell, suggested in a humorous letter to his friend Lord Kelvin that this had happened because Thomson was 'bringing his obtrusive science to bear upon the engineers, so they broke the cable with not following (it appears) his advice'. Maxwell added the words of a song that he had composed, named 'The Song of the Atlantic Telegraph Company'. The first two verses went:

2(U)
Mark how the telegraph motions to me,
2(U)
Signals are coming along,
With a wag, wag, wag;
The telegraph needle is vibrating free,
And every vibration is telling me
How they drag, drag, drag,
The telegraph cable along.
2(U)
No little signals are coming to me,
2(U)
Something has surely gone wrong,
And it's broke, broke, broke;
What is the cause of it does not transpire,
But something has broken the telegraph wire
With a stroke, stroke, stroke,
Or else they've been pulling too strong.

Here, '2(U)', Maxwell indicated, was shorthand for singing 'Under the sea' twice, making his ditty a precursor to the song of the same name in Disney's *The Little Mermaid* (though it's unlikely that Howard Ashman, who wrote the Disney lyrics, was influenced by Maxwell's work).

Resistance is futile

One significant issue that Thomson and another of Maxwell's friends, the English engineer Henry Fleeming Jenkin, faced was overcoming the electrical resistance of the extremely long cable. Resistance (R) is one of the key measures in an electrical circuit, which, you may remember from school, links the voltage (V) and the current flowing (I) with the simple formula $V = I \times R$.

Thomson, with some help from Michael Faraday, worked out that with the resistance that would arise from the original specification for the cable, it would mean that it could take as long as four seconds to send a single character. At this stage, electrical resistance was poorly understood. To help in part with the transatlantic cable project, Thomson, Maxwell, Jenkin and another Scottish physicist, Balfour Stewart,* were given the task of pinning down an appropriate electrical unit for resistance.

Thomson devised a surprising mechanism to do this. It involved rapidly rotating a coil of wire. Because this was taking place in the Earth's magnetic field, it would induce an electromagnetic effect, which was balanced out against the Earth's magnetic pull on a small permanent magnet. Because both effects involved the Earth's magnetic field strength, this cancelled out, meaning that only the size of the coil, its speed of rotation and its electrical resistance would influence the degree to which the permanent magnet was deflected.

As a result of its design, Thomson's apparatus measured electrical resistance as a velocity. This caused lots of confusion on the transatlantic cable project, as many thought it referred to the speed at which a signal could be sent down the wire. A rate of 10 million metres per second was defined as the standard unit of resistance, which was initially given the name 'ohmad' to make it similar to the farad, the unit of electrical capacitance already named after Michael Faraday. This was simplified to 'ohm', both words based on the surname of the German physicist Georg Ohm, who had established the relationship between electrical resistance and current.

The cable was completed in 1858 with great celebrations on both sides of the Atlantic, but before long there were more problems. Over a short time, the signal became weaker and weaker. Whitehouse suggested increasing the voltage used to 2,000 volts, without success. Signals stopped flowing entirely within three weeks, which it has been suggested was primarily due to overloading from the excessive voltage. Thomson and a number of other scientists were tasked with analysing the problem and fixing it.

---

* Balfour Stewart sounds like the surname of the Scottish physicist, but it was his full name, landed with a first name of Balfour. It is not recorded if he was known to his friends as Balfy.

With changes in design and signalling apparatus, to overcome the first cable's limitations Thomson was able to persuade Field to invest a further $2.5 million in a second cable – that's $96 million now with inflation and the equivalent of a construction project costing around $1.5 billion. This cable also broke, leading to a third cable being laid in 1866, which was finally successful. This was joined by a second working cable when the broken one was repaired soon after. Such was the value of the communication facility that, despite the huge outlay, Field had earned his investment back in just three years. This was helped by a hefty $1 per character charge for sending a message. In inflation terms alone that's around $38 per letter and compared to current salaries around $550.

While electricity was beginning to be used practically, another key role of the electron in nature was coming to the fore, though its involvement was not realised at the time. The chemical elements have distinctive behaviours as a result of the electrons that surround them – and a combination of similarities between these behaviours in some elements and their atomic weights would be used to construct a table that helped turn chemistry into a proper science. We need to backtrack a few decades to find a Russian scientist at work, unwittingly reflecting a very different role for the electron.

MR M'S TABLE 10

Even though Mendeleev always denied that electrons exist, they later turned out to be vital for ordering the elements in his table.

Patricia Fara in *Science, A Four Thousand Year History* (2009)

It's an inevitable aspect when covering the life of something as long-lived as electrons that there will be instances where they played an important part in history without it being recognised at the time. A lot of our focus so far has been on free electrons, able to move around as an electrical current or to build up as a static electrical charge. But as far as matter is concerned (including the matter than makes up every human being), it is the electrons bound in chemical elements that make all the difference. Chemistry is a science that is run by electrons.

Black box events

Science often involves what might be called black box events. Imagine you have a box with a number of openings. You can drop a ball bearing or a marble into any one of the holes and it will emerge from another. All we can do is observe and measure what occurs on the outside of the box. We can try dropping balls into different openings, but we can't look inside the box.

Scientists will then build a mental picture of the inside of the box, known as a model. This could be, for example, as simple as working out what would happen if there were a set of partitions inside the box, or there could be a complex mechanism that changes the direction of the balls as they pass through – or even a living creature inside that pushes each ball whichever way it fancies. After developing a hypothesis, scientists might then devise new experiments – for example, what happens with different-sized balls, or if we hold the box at different angles, or even perhaps immersing the box in a liquid to see if the model holds up, or it needs changing.

Atoms had been speculated about since ancient Greek times, when some philosophers offered the idea that matter could be cut up into smaller and smaller pieces until you got to an uncuttable (atomos) piece. This was an alternative to the more dominant idea of the four Earthly elements thought to make up all familiar matter: earth, air, fire and water. Around the start of the nineteenth century, English chemist John Dalton had proposed that the chemical elements, the basic building blocks of matter, were made up of atoms, indivisible particles which always combined in the same whole-number ratios to make compounds. However, his basic atomic table – which included a number of chemicals such as lime (calcium oxide) that weren't atoms – was based simply on weight and did not give any feel for a pattern of behaviour that linked similar elements.

In the nineteenth century, atoms were entirely black boxes,* but scientists could look for patterns and try to build up a logical structure that would model how they behaved. When the periodic table of the elements was devised, that iconic image so familiar from the walls of school chemistry labs, the patterns they were observing were entirely the result of the actions of electrons, but our biographical subject's role was hidden. These patterns seemed to reflect an unknown structure within the black box of each element which meant that, for example, the metals sodium and potassium both reacted in a similar dramatic way with water.

---

* Atoms were black boxes at best, as their very existence would still be questioned by some scientists as late as the early twentieth century.

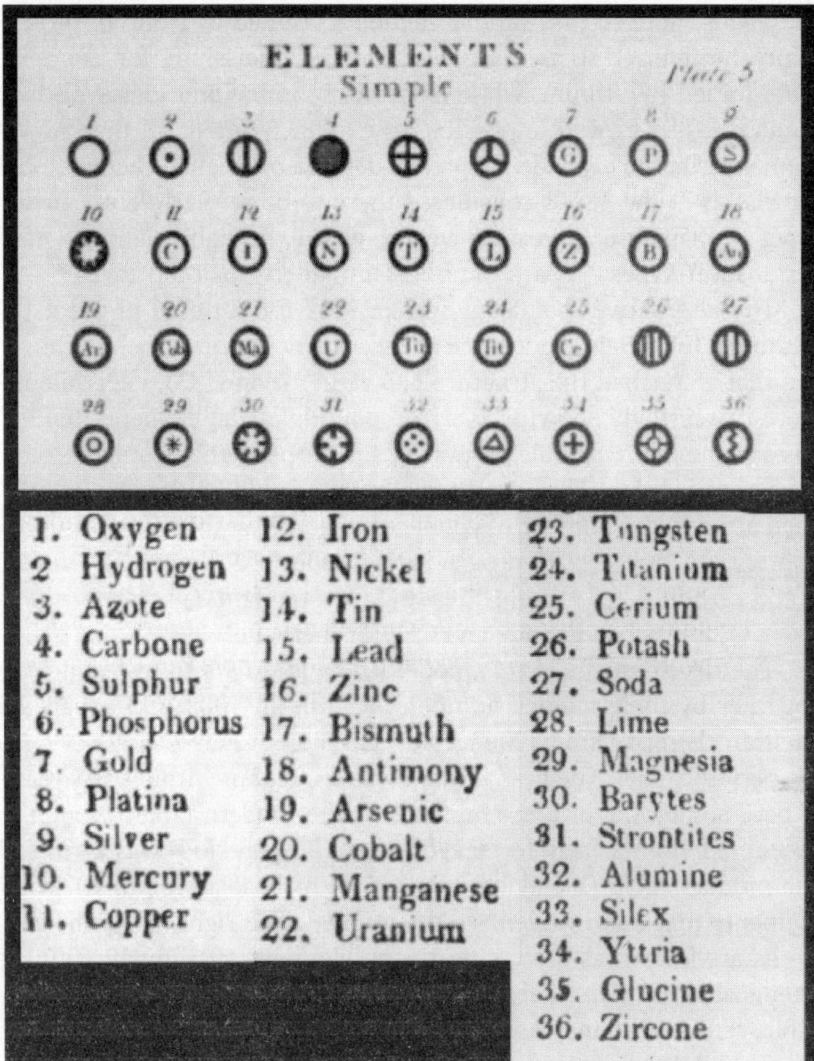

Dalton's table of elements.

Searching for patterns

Among the first attempts to sort out an overall structure for elements was made by German chemist Johann Döbereiner who, as early as the 1820s, noticed that similar elements seemed to group in threes, which he referred to as triads. Sodium and potassium, for example, were joined by lithium, while chlorine, bromine and iodine all had similarities. It proved impossible to extend triads to all the known elements, but an expanded approach devised by English chemist John Newlands in the 1860s considered there to be seven different broad types of element of increasing weight, where the eighth started a new set of seven types – a pattern not dissimilar to musical octaves.*

The alternative to analogy was to look for patterns in the data. Although John Dalton got some of the specifics wrong (such as thinking that water had the structure HO rather than H_2O), experiments were increasingly showing that compounds of the elements tended to combine in predictable proportions, a property that would become known as valence. These valence numbers could be thought of as if each atom had a number of connectors, enabling them to join up in appropriate multiples. Hydrogen has a valence of 1, oxygen 2, nitrogen 3 and carbon 4, for example, leading to some familiar combinations like CO_2 for carbon dioxide and CH_4 for methane.

The development of this model led to attempts to collect atoms together by their valence numbers, leading in 1869 to 35-year-old Russian chemist Dimitri Mendeleev, based in St Petersburg, devising his periodic table. The first attempt was crude, but within a few years it bore something of a resemblance to the modern table (though it contained significantly fewer elements). The 'period' was a row of the modern table (Mendeleev started with these as columns), with elements that vary in valence sequentially left to right, while the columns are known as groups, featuring elements that exhibit similar chemical behaviours. Each step along a period, or further down the table, reflected an increase in the relative weight of the atom.

By building this structure, Mendeleev was able to point out a number of gaps in the known tally of elements. At the time, 63 elements had been identified (compared with the current 118). Mendeleev gave

* It might seem bizarre to base a pattern of elements on musical notes, but scientific models are often initially based on analogies. Isaac Newton is thought to have come up with (an entirely arbitrary) seven colours of the rainbow using a similar musical analogy.

his proposed extra elements names beginning with 'eka' (Sanskrit for 'one') followed by the name of the element immediately above in the same group. For example, immediately below silicon came ekasilicon, filling one of the gaps in the fourth period. He was able to identify what several of its properties would be before what's now known as germanium was discovered seventeen years later.

Unfortunately for Mendeleev, while his neat pattern worked neatly for periods 2 and 3, where there were complete runs of the eight possible groups, the rest of the periods were inconsistent in their structure. Period 1 initially contained just one element (hydrogen), which would be joined by helium after its discovery in the 1890s. Periods 4 and 5 had an extra set of eight groups, while 6 and 7 would eventually be found to have an additional set of fourteen groups. The underlying pattern was anything but simple – and the explanation for this structure would have to wait for an understanding of the electron's role (it will have to wait here for chapter 24).

Clearly something about the nature of atoms lay behind the patterns observed in the periodic table, and there would be much speculation to come on one of the electron's most remarkable creations. Mendeleev, unfortunately, was no enthusiast for the electron once the concept had been introduced, decades after his development of the periodic table. In the 1905 edition of his seminal textbook *The Principles of Chemistry* he noted that as far as he was concerned the 'hypothesis of electrons' was 'scarcely conceivable'. He wrote, 'It appears to me that the whole question of a primary matter belongs to the province of fancy and not science.'

For now, though, back in the 1870s, with the electron itself still hiding shyly away, practical developments arising from the work of Faraday and his contemporaries would lead to the first conscious deployment of electricity that went beyond a means of communication or entertainment. Two possible applications – motors and lighting – were neck and neck in development, but one just made it into practical use first. It was time to put the electron to work.

11 THE ELECTRIC HORSE

The simplicity attached to the operation of an electric vehicle by any person of ordinary intelligence is too well known to need comment at this point; but it is found from experience that there is the same difference in the care taken of an electric vehicle that there is among men who attend dynamos and steam engines, or drive horses, with a corresponding difference in troubles and aggravations.

Clinton Edgar Woods, quoted in *The Evolution from Horse to Automobile: A Comparative International Study*, Imes Chui (2006)

Take a stroll along Madeira Drive on Brighton's seafront, and opposite the Sea Life Centre you will come to an unimposing modern building labelled Volk's Electric Railway. Here it's possible to see in action (or take a ride on) the oldest public implementation of the electron as motive power for a vehicle that's still operating today. The 1.6-kilometre (1-mile) track has been carrying passengers in its small electric trains, powered from a third rail, since 1883.

Initially a 2-foot gauge track (compared with the now standard railway gauge of 4 foot, 8.5 inches – or 1.435 metres) was used, with the 50-volt DC electricity provided through the running rails. This would be upgraded to 2 foot, 8.5 inches (0.825 metres) at 160 volts in 1884 to cope with an increase in track length, with the power moved to a separate third rail in 1886, as this was easier to insulate from earth, reducing leakage.

When electrical engineer Magnus Volk opened his track, it was a novelty, but it was not the first time that electricity had been brought into use as a means of motive power after the development of the electric motor following Faraday's work. Because of the limitations of batteries at the time, the only lasting use of electric vehicles (including Volk's) was initially for railways, trams (trolleys) and trolley buses, which could pick up their electric power from rails or wires as they travelled along.

The first steps

The earliest electric train on record appears to have been an experimental implementation on a 1-kilometre (0.62-mile) stretch of the Miller's Line that ran in Russia from Beloostrov to Sestroretsk off and on for about ten years from 1875. Other experiments were undertaken, notably by leading German electrical engineer Werner Siemens, who first demonstrated an electric passenger train at the Berlin Trade Exhibition of 1879 after working on electric motors and generators for over ten years. But the first truly commercial application of electric vehicles was the Gross Lichterfelde Tramway on the outskirts of Berlin in 1881. Also built by Siemens, the track ran for about 2.4 kilometres (1.5 miles), using 180-volt DC motors, like the original Volk's Railway powered through the running lines.

Unlike Volk's track, the Gross Lichterfelde trams were not separated entirely from road transport and human feet. Although attempts were made to protect the public by either having short dead stretches of rail that the cars could coast across, or switching so they were only powered when a tram was about to cross, accidental electric shocks were not uncommon. This led to a further development as early as 1883.

Where some railways would continue with live rails until the present, Siemens realised that this approach was not ideal for tramways and demonstrated the first overhead cables for powering vehicles at the 1881 Exposition of Electricity in Paris.* This would be first put

* It was a mark of the growing commercial importance of electricity that 1881 would see a trade fair, the International Exposition of Electricity, dedicated to it and its applications from motors and lighting to the telephone. We will see in chapter 15 how central electricity was to the Great Columbian Exposition of 1893.

into action in a live tramway in 1883, reaching Gross Lichterfelde in 1893. Sadly, the trams ceased to run in 1931.

Leaving the tracks behind

Railways and tramways are limited by their tracks. It can be more useful if a vehicle can follow more flexible routes. Siemens would experiment with an electric bus (what would become known as a trolleybus) in 1882, and for some time these provided a useful addition to city centres, where they could be deployed with just the addition of overhead wires, replacing the need for expensive and obtrusive rails.

I can remember being taken as a young child on the trolleybuses in Manchester, but they were phased out in 1966. Diesel buses were considered a more practical replacement for the disappearing trams in many UK cities, having total flexibility of route, though ironically trams have made a significant comeback in recent years, in part due to the aim to move away from fossil fuels.

While electricity would continue to be an important motive power for trains, thanks to their fixed routes and dedicated spaces – and have become even more dominant since steam was phased out, with diesel now also being replaced – on the roads, particularly for individuals, petrol and diesel cars would dominate for over a hundred years.

There were indeed early electric cars, but the limitations of the battery technology of the day soon became apparent. The problem is one of energy density. Fossil fuels like petrol and the kerosene used as aviation fuel pack considerably more energy into the same amount of space as even the best modern battery, typically around 20 times that of a lithium-ion battery. In the early days of electric vehicles, as much as 100 times as much energy could be stored in the same volume.*

Electric road vehicles have existed since the early days of motorised transport, but they were limited for many decades to short-range

* Energy density is one reason we are unlikely to see hydrogen cars in the mainstream. Even liquid hydrogen only has a third of the energy density of petrol, while highly compressed gas limps in at around one seventh of the energy.

vehicles, such that as late as the 1980s the only electric vehicles typically seen on the roads of Britain were very slow, short-range milk floats. However, with the development of lithium-ion batteries we are finally seeing an ability for electric cars, buses and lorries to take to the roads.

But electric motors did not represent the only opportunity for the generator to bring the power of the electron to practical use. One obvious possibility was to use electricity for lighting to replace the crude, flame-based lighting that had dominated since ancient times. However, it proved a harder challenge than many anticipated.

LET THERE BE LIGHT 12

I speak without exaggeration when I say that I have constructed 3,000 different theories in connection with the electric light, each one of them reasonable and apparently likely to be true. Yet only in two cases did my experiments prove the truth of my theory.

Thomas Edison, quoted in *Talks with Edison (Harper's Magazine)* by George Parsons Lathrop (1890)

Natural light is wonderful but has obvious limitations. It can vary hugely (a full moon, for example, is around 300,000 times weaker than bright sunlight), limits nighttime activity and is of limited value in enclosed spaces. Ever since the discovery of fire at some unknown date in the past, probably due originally to a natural source such as a lightning strike, the flame had provided the best option for nighttime lighting.

The move had been gradually made from fires to oil lamps and then to gas lighting, enhanced by the use of mantles, uninflammable materials that provide a bright uniform glow when heated by a flame, which would have been state of the art when Faraday was delivering his lectures on electricity.* But even though gas lighting would remain

* One of Faraday's best-known contributions to the Christmas lectures series at the Royal Institution that he founded was *The Chemical History of a Candle*, given in 1848, where he explored the nature of the ubiquitous flame that his work on electricity would render more decorative than practical.

in use for a surprisingly long time in some locations (when I went to school by train in the late 1960s and early 1970s, Rochdale station still had gas lights), it was problematic.

Gas lighting gave off more heat than light, was not particularly bright and the presence of flames from gas lights indoors in everything from private homes to theatres, offices and factories was a constant fire risk. Even the functioning of such fire-based lights is down to the action of the ever-present electron. The heat of the chemical reaction in a flame stimulates electrons around the atoms involved, pushing them up to higher levels around the atom. As they drop back, they give off energy in the form of photons of light. But electric lighting would provide an opportunity to take out the middleman in the process.

Straight out of the arc

Faraday's mentor, colleague and occasional opponent Humphry Davy was an early player in the attempt to harness electricity for lighting, using one of Faraday's generators to heat up a platinum wire until it glowed. The process here is electrical resistance. As electrons begin to move under the influence of an electrical potential they hit atoms in the conductor, converting kinetic energy into heat. This is double electron action – both generating heat as they move and giving off light when stimulated in atoms. But the problem with this approach to lighting is that if wires are heated enough to glow brightly, they quickly melt or burn away.

Davy had had more success experimentally with electric arcs, which he had developed somewhere between 1801 and 1809, before Faraday's electrical successes. Exact details are sparse, but Davy seems to have made a first attempt on moving to the Royal Institution in 1801. He then made use of large batteries of voltaic piles in chemical work, before returning to the arc light for a dramatic demonstration in 1809.

This arc light consisted of a pair of charcoal rods, initially with a narrow gap between them, described in one source* as 'within the

* The source describing the detail of Davy's arc lamp is *Electrical Engineering for Electric Light Artisans*, William Slingo and Arthur Brooker (1900).

thirtieth or fortieth part of an inch' – between 0.64 and 0.84 millimetres. To produce the required voltage, he employed an electrical battery consisting of 2,000 cells. Bear in mind this was within ten years of Volta first devising the technology. Once the arc was struck, the charcoal pieces were gradually separated until they were about 4 inches (10 cm) apart. Davy's vast battery raised the voltage sufficiently to exceed the point at which air molecules ionise, causing a flow of electricity across an air gap (around 30,000 volts per centimetre). Those who saw his light claimed it rivalled the Sun for brightness.

Because the spark was curved, according to the description of the experiment from 1900, Davy described 'a most brilliant ascending arch of light, broad and conical in form in the middle'. The authors Slingo and Brooker note that with vertical electrodes the resultant light would 'bear little or no resemblance to an arch. The term arch, in its abbreviated form arc, is, however, retained as the name of the luminous space between the carbons.' This seems based on a misunderstanding of the origin of the word arc: Davy may have commented on the resemblance to the architectural feature, but the mathematical 'arc' would have been the more appropriate scientific term.

It wasn't until generators were available that arc lights could be more than a demonstration – batteries did not store enough energy to run them for any length of time. The first installation is generally thought to have been at the South Foreland lighthouse at Dover in 1858.* This was at the instigation of Michael Faraday, who acted as scientific advisor to Trinity House, the body that looks after UK lighthouses. The arc lighting was trialled there for two years, then first set up permanently at Dungeness, also in Kent, with Souter Lighthouse in Sunderland being the first designed specifically for electric lighting.

Arc lights would continue to be used where extremely intense lighting was required for decades – but they aren't suitable for most everyday use, as they are uncomfortably bright, require dangerously high voltages and can easily result in fires. As well as searchlights,

* South Foreland lighthouse would have another passing role in the biography of the electron, being used as the first UK station for a radio transmission across the English Channel from Wimereux in France in a Marconi demonstration in 1899. See chapter 14 for more on Marconi's work.

one of the last uses was in cinema projectors through to the 1960s, despite the obvious potential for disaster of using this approach with flammable film stock.

Fishing for a filament

Despite much effort, getting an electrical light to work using a metal filament without the filament burning out continued to be problematic. For a brief period, it seemed likely that a mechanism similar to that of Crookes tubes would be used. Lights based on this principle, pioneered by German physicist Heinrich Geissler since the 1850s, made use of the glow within a low pressure sealed tube when a voltage was placed across it, and would lead to various kinds of practical discharge tubes, including both fluorescent lighting and neon lights – but at the time it was not possible to get tubes bright enough for practical use. (We'll find out more about Crookes tubes for a different reason in chapter 16.)

It would be in 1879 that things changed, almost simultaneously on both sides of the Atlantic. Just as had been the case with the telegraph system, there was a transatlantic race for a successful implementation, this time between American inventor Thomas Edison and English inventor Joseph Swan. Unlike Davy, both would move from metal filaments to carbon, with what perhaps should have been an obvious mechanism to prevent the filaments burning out.

Edison is more of a mythical character than most who would be involved in the electron's relationship with humanity. Born in 1847 in Milan, Ohio, he had limited schooling and by the age of twelve was working as a train boy on the Grand Trunk Railway. Unlike others in this role, he seems to have spent as much time experimenting as doing his job, initially in a mobile lab on the train, then, after a brief stint in the basement, moving his experiments (with permission from the company) to an empty freight car.

By the age of fifteen he was producing a newspaper from the lab which he sold to passengers on the train. Then came his breakthrough moment, certainly according to the legend that built up around Edison. The son of the stationmaster where Edison was due to pick up his next service slipped and fell in front of an oncoming train. Edison jumped onto the track and pulled the boy clear, saving his life. As a reward, the stationmaster showed Edison how to use the electric telegraph, inspiring his electrical experiments.

Given the existence of the lab and all the evidence of Edison's drive to succeed, it seems highly unlikely that such an occurrence would have been necessary before he could get his hands on a Morse key. The rescue may have happened, but it's unlikely to have been a factor in inspiring Edison's inventiveness. Before long Edison's company (now moved out of the freight car) was churning out inventions. Not only the light bulb, but everything from the phonograph and an early form of moving pictures to the 'electric pen'.

Edison employed his legendary patience in trying different possibilities that might make an electric light bulb possible – and the result was certainly a success. By the end of 1879 he had readied a demonstration to win over investors. Press and New York city officials took some time away from New Year's Eve celebrations to see electric lighting in action at Edison's Menlo Park facility in New Jersey. It proved to be a perfect evening for it – the grounds of the lab were covered in snow, which set off perfectly the 290 bulbs that Edison switched on, a magic illumination that was made even more dramatic when he was able to start them flashing.

Getting there first

Despite his legendary successes, Edison was not the first to invent a working light bulb. It was Joseph Swan who beat him to it. Born in Sunderland in the North East of England in 1828 (so, nineteen years older than Edison), Swan started work as an apprentice pharmacist. By the time he was eighteen, Swan was offered a partnership in a chemical manufacturing company. Just four years later (when Edison was three years old), he began experimenting in his home laboratory with electric lighting. It took him ten years to get a working, short-lived bulb.

At the time, getting a good enough vacuum to avoid the filament burning out quickly was an obstacle, but Swan returned to the problem in 1875 and three years later had a working bulb, which he was able to publicly exhibit by February 1879. Swan had demonstrated his bulb eight months earlier than Edison, but had not at that point obtained a patent. With his trademark cutthroat approach, despite being perfectly aware that Swan had beaten him to it, Edison sued for patent infringement.

Patent law sometimes fails to protect the earliest inventor – but in this case Swan triumphed, establishing his prior claim. As part of the settlement, the Edison and Swan United Electric Light Company was set up to sell what would become known as the incandescent light bulb. The response of the then dominant gas lighting companies doomed them. Rather than move into electrics, using their scale to buy out the newcomers, they thought they would stay on top by bringing out bigger and better gas lights. Soon they had gone.

The earliest experiments that led to the discovery of oxygen had shown that a lack of it in an enclosed space made it impossible for materials to burn. Both Edison and Swan realised that taking the oxygen out of a glass bulb made it far easier to keep a white-hot glowing filament intact – though it still took experiments with many different filament materials and designs before each got one to work. These were initially carbon (Edison produced his from carbonised bamboo) but would go on to be made of metals such as osmium and tungsten.*

The battle for supply

Significant though the development of the electric light bulb was, it's arguable that its most lasting benefit would be the supporting infrastructure needed to operate it. Up to this point, electricity had largely been supplied first from Leyden jars and then from batteries based on Volta piles. But lighting buildings or streets on batteries alone was not practical. Electric lighting would demand wiring systems to deliver the requisite electrons – and central generators to provide a constant

---

* The metal elements osmium and tungsten used in light bulb filaments were responsible for the name of what is probably the best-known German electric lighting company, still in operation, Osram. Osmium is clearly responsible for the 'Os' part, but the 'ram' is more puzzling unless you are aware that in German, tungsten is known as wolfram, hence its chemical symbol W.

flow. It would be electricity supply that would give the electron a role in establishing large corporations – and lead to a pitched battle over the means of supply.

Until the late 1800s there had been little question over the way that electrons were provided for use to potential customers. Even the names associated with electricity such as 'current' suggested a simple flow of electricity from one pole of the battery (or other source) to the other. This approach would become known as 'direct current' (DC), to contrast it with 'alternating current' (AC), where the electricity flowed first in one, then in the other direction, switching many times a second.

This second approach at first sight seems crazy, introducing an unnecessary complication, but as we will see it had significant advantages. It was championed by the controversial Serbian-American engineer Nikola Tesla (whom we'll meet again in a couple of chapters). Tesla was born in Smiljan, then part of the Austrian Empire, but now in Croatia. He attended the Austrian Polytechnic in Graz, but dropped out without a degree – he is said at this stage in his life to have had more interest in gambling and socialising than advancing his education. Four years later, he got a job in Paris as an electrical engineer with the Continental Edison company, moving to New York with the company in 1884.

Although Tesla was a rising star in Edison's labs, the two men's attitude to life was very different. This came to a head when Edison is alleged to have told Tesla that, should he successfully improve the company's midsize generator, he would give his young protégé a bonus of $50,000. That's an enormous amount – equivalent to around $1.7 million allowing for inflation, or around $16 million when comparing incomes between then and now. Tesla did the job, but Edison is said to have refused to pay up, claiming it was a humorous remark and showed that Tesla didn't understand American humour. Within a year, Tesla had left Edison to set up his own company, Tesla Electric Light and Manufacturing.

AC has a number of potential advantages over its more straightforward cousin. Most importantly, it is far easier to modify voltage levels. High voltage is better for transmitting electricity, as there is less loss to heat. When you pass electricity down a wire, the power lost to heat is described by the simple formula I^2R – the square of the current times the electrical resistance of the wire. If the resistance or the current is high enough, the loss becomes considerable and the wire will start to glow – this is how both electric fires and incandescent light bulbs work.*

The last thing you want when transmitting electricity is to lose a lot of your power to heat. This was a real issue for Edison with his low voltage (110 volts) DC system – he could only usefully transmit power over a distance of around 2.5 kilometres (1.5 miles). But since power is based on current multiplied by voltage, if you increase the voltage significantly, you can transmit the same amount of power using a lot less current. And bearing in mind that the heat loss is proportional to the square of the current, that vastly reduces loss. Modern power lines can run at as high as 400,000 volts, specifically to minimise loss.

It was soon realised that high voltages would make long-distance provision of electricity possible, but the supply still needed to be stepped down to a more manageable level for practical use. A number of options were tried as a standard voltage, with most of the world opting for between 220 and 240 volts, but the US stayed close to Edison's original at 120 volts. Bearing in mind that Faraday's electrical induction process works by producing a current in a second coil as electrical fields rise and fall from a primary coil, it proved very easy to switch voltages when using AC by using induction between two unequally wound coils in what would become known as a transformer.

The AC/DC finale

For Edison, direct current, which was easier to generate and the obvious way to drive electric motors, was the obvious solution. It wasn't until 1887 that Tesla had been able to develop the first effective AC motor: it was painfully ironic that the rival to Edison's system would use technology developed by a man whose career he had kickstarted.

* The fact that an incandescent light bulb glows brightly indicates it is losing a lot of energy to heat. One of their biggest problems was they were essentially heaters that also glowed a bit: around 90 per cent of the power went to heat, not light.

Tesla was dismissive of Edison's painstaking approach, writing, 'If Edison had a needle to find in a haystack, he would proceed at once with the diligence of a bee to examine straw after straw until he found the object of his search ... I was a sorry witness to such doings, knowing that a little theory and calculation would have saved him 90 per cent of his labor.'*

While Edison stuck stubbornly to DC, millionaire developer of railway technology George Westinghouse embraced the superior AC technology. After some initial investigations, Westinghouse bought into Tesla's designs. Arguably, the real breakthrough for Westinghouse (and for Tesla) came at the 1893 World's Columbian Exposition in Chicago.

Ever since Prince Albert had organised the Great Exhibition in London in 1851, countries had put on 'World's Fairs' where exhibitors from around the world would come to show off their wares and technology. Designed to commemorate the 400th anniversary of Columbus' arrival in the New World, the World's Columbian Exposition was one of the greatest examples of its kind ever held.

This was a cross between a trade fair and a theme park, based on a huge 600-acre site. There were plenty of worthy demonstrations of machinery and agricultural technology – but there was also a half-mile boulevard named the Midway Plaisance.† This featured impressive entertainments. The first ever Ferris wheel, 80 metres (264 feet) high, gave riders the thrill of a lifetime. There was a panorama of the volcano Kilauea, a Lapland village, a tethered balloon ride, an ice railway, a Japanese bazaar, a Venetian glassworks and a scale model of St Peter's in Rome. Most popular of all, bringing in over 2.5 million visitors during the six-month run of the Exposition, was A Street in Cairo.

This included camel rides and a bazaar, but it became notorious thanks to its exhibition of *danse du ventre*, which would become known as belly dancing or hoochy-coochy dancing. The show was described in the *Chicago Tribune* as a 'depraved and immoral exhibition', ensuring its

---

* Tesla's criticism of Edison's technique comes from a piece he wrote for the *New York Times* in 1931.

† The main axis of American carnivals are still often called the midway, inspired by the Midway Plaisance at the Great Columbian Exposition.

huge popularity. Nearby was a technological marvel, the strange sounding Zoopraxographical Hall, which was the first ever purpose-built cinema, where photographic pioneer Eadweard Muybridge showed off his early moving pictures. And all this excitement demanded something equally novel on the lighting front.

The fair's organisers were looking for a supplier of power both for a dedicated Electricity Building and to provide lighting across the park. Edison's General Electric company quoted $1.7 million to do the job with DC. Tesla, with the backing of Westinghouse, offered to provide an AC system for about one third of the cost. Edison undercut this at $554,000 – but Westinghouse sealed the deal at a bargain $399,000. Edison was a far better businessman than Tesla would ever be – but Tesla time and again would demonstrate his abilities as a showman. His 93,000 electric bulbs not only provided light, but outlined the buildings, making the site a standout phenomenon at night. The fair was using three times as much electricity as the entire city of Chicago.

This was the beginning of the end for domestic DC: Edison would lose the battle of the currents. He attempted to fight back commercially, eventually resorting, according to popular legend, to a bizarre stunt. In 1903, a funfair called Luna Park had been set up on Coney Island in New York. One of their attractions was to be Topsy the elephant, who had been bought from a circus after she killed a member of the audience. Finding Topsy difficult to control, the owners of Luna Park decided to execute the elephant as a publicity stunt, originally intending to hang her.

After complaints from the American Society for the Prevention of Cruelty to Animals, it was decided that Topsy would be killed in a supposedly more humane manner. She was to be simultaneously strangled by winch, poisoned with cyanide (painted onto carrots) and electrocuted. The event was filmed using early movie cameras by the Edison Manufacturing Company, who also arranged the technology for the electrocution. But rather than use their own DC supply, Edison's company used 6,600 volts of AC, claiming that this would demonstrate the risks of using AC and that (with no good evidence) the ability to kill an elephant proved that AC was more dangerous than the Edison alternative.

In reality, by 1903, AC had already won the day – and this whole story is more myth than history. The electrocuted elephant was never employed as a stunt to prove the advantages of DC. It appears to have been mixed up with a series of electrocutions of animals (not including an elephant, but going as large as a horse) undertaken around 1888 by an electrical engineer called Harold Brown, who arguably correctly said that lighting companies using high-voltage AC were responsible for unnecessary fatalities and wanted (with the support of Edison's company) to demonstrate the dangers of this. This was accompanied by a deliberate push to make sure that AC was used for the electric chair, introduced for executions in the late 1880s. Tesla would correctly protest that AC, correctly handled, was safer than DC, but there was no doubting Edison's determination to win.

Domestic supply is now universally provided as AC, with voltages stepped up to hundreds of thousands of volts when carrying it on long-distance cables. The early electric trains were DC, but now all modern trains use higher voltage AC. Despite this, DC has not gone away. All battery-operated devices make use of DC – and even high-voltage DC is making something of a comeback, as it has proved that this can be used to transport long-distance power with fewer losses now that we have the technology to step down DC voltage as well as AC.

The discharge glow

The basic approach of heating up a strip of metal in a low-pressure environment that prevented it burning up straight away would remain the standard approach for everyday lighting in homes for over a hundred years. However, the earlier approach of producing a glow from an electrical discharge across an evacuated tube did not go away. It took a significant step forward when it was realised that the power of such tubes could be enhanced using fluorescent materials.

Gas-discharge lights produce a lot of high-energy ultraviolet light that is not visible (and has the potential to damage eyesight). However, if the outside of the tube is coated with a phosphor, a material that absorbs ultraviolet and gives off light in the visible range, it becomes a more efficient way of turning electricity into light than an incandescent bulb. Such fluorescent lights were more common in businesses and schools than homes, and their harsh, unnatural lighting was considered second rate to the warm glow of the traditional bulb.

In the last decades of the twentieth century, though, the limitations of the ageing technology of incandescent lights were starting to become apparent. As we have seen, an incandescent bulb uses more of the energy it derives from a flow of electrons to produce heat than it does light. This was always wasteful – but with growing awareness of the importance of combating climate change, that unnecessary expenditure of energy still derived in large part from burning fossil fuels became increasingly embarrassing.

Over the years, fluorescent tubes had become more versatile, and there was a short-term advantage to be gained by switching to bulbs that used a small, coiled or looped fluorescent tube, often sacrificing some efficiency to produce a warmer glow. But these bulbs took minutes to achieve full brightness, and only reduced the energy consumption by a factor of three or four. There had to be something better – and there was. But the technology had a major flaw.

The cat's whiskers

The groundwork for the approach that would transform electrical lighting at the end of the twentieth century had already been laid just a few years into it. In 1907, the English electrical engineer Henry Round was experimenting with so-called cat's whiskers. These had nothing to do with feline hair – they were junctions between a wire and a semiconducting material such as lead sulphide that allowed current to flow only in one direction (a solid-state diode) and were used in early radio receivers, to separate the signal from an underlying 'carrier' wave.

Round noticed that some of his experimental cat's whiskers glowed when in use. He wrote to the magazine *Electrical World*: 'On applying a potential of 10 volts between two points on a crystal of carborundum [silicon carbide], the crystal gave out a yellowish light. Only one or two specimens could be found which gave a bright glow on such a low voltage, but with 110 volts a large number could be found to glow.'

The phenomenon was given the name electroluminescence, but the mechanism behind it was not explained. It would require an understanding of quantum physics to realise that the glow was the result of energetic electrons in the so-called conduction band (more on this in chapter 31) dropping down into a lower energy gap (a hole), with the energy that the electrons lost released as photons of light. Though a few people experimented with these interesting light sources, it wasn't until the 1960s that anything practical would emerge. As we have seen, an electronic component like the cat's whisker that only allows current to flow one way is known as a diode,* so this new type of product became known as a 'light emitting diode' or LED.

The first LED was patented in 1962 by James Biard and Gary Pittman at the US chip maker Texas Instruments. This component worked in the infrared – not quite visible light, but useful for optical communications devices. A year later, Nick Holonyak Jr. at General Electric had produced the kind of LED that dominated for some time – the red LED, which would initially be used as indicator lights and later in basic LED displays. LEDs were very appealing because they used a tiny amount of energy to produce a glow – so were ideal for small, battery-powered devices like calculators and watches – and produced very little heat.

At the time there was no prospect of using LEDs for functional lighting. They were bright for their size, but nowhere near bright enough to light a room – and even if they could, there were limited uses for purely red lighting. What was needed was either a white LED or both green and blue ones that could be combined with red to produce white light. Green was soon available, but the real challenge proved to be blue, which is why three Japanese scientists – Isamu Akasaki, Hiroshi Amano and Shuji Nakamura – would be awarded the 2014 Nobel Prize in Physics for the development of a commercially usable blue LED.

Using LEDs in a red-green-blue combo would prove useful for TVs and colour-changing lightbulbs, but was too clumsy and expensive

* We will find out more about diodes – and where this name comes from – in chapter 23.

for mass-market white LEDs*. Most LED lighting now makes use of a single intense blue LED source which is coated with a yellow phosphor – a material that acts similarly to the coating of a fluorescent light, absorbing much of the blue light and transforming enough of it to red and green to produce an overall white output.

LED bulbs now seem so commonplace that it is easy to overlook just how stunning a change they made to lighting efficiency. Not only do they last up to a hundred times longer than incandescent bulbs, whose filaments gradually lose metal until they burn out, but their ability to convert a flow of electrons into light is an order of magnitude greater. Where a traditional bulb would convert around 4 per cent of the electrons' energy into light, a white LED will make use of more than 50 per cent. This is a remarkable improvement, thanks directly to the actions of electrons.

If a major gift from the electron to the world was the ability to easily and safely lighten the darkness, another was to offer a new, near-instantaneous mechanism to communicate around the world. The telegraph had started the communication revolution, but for ordinary people it would be the ability to send voices down the wire that would make all the difference.

* From school, you may remember being told that the primary colours are red, yellow and blue. They aren't, they are red, green and blue – from these coloured lights any other colour can be produced. The secondary colours, cyan, yellow and magenta, are used in pigments to react to the whole range of light colours, and these are simplified to blue, yellow and red for children

13 VOICES DOWN THE WIRE

I went [to the Centennial Exposition of 1876], and when I returned Professor Tingley's first question was, 'Did you see the telephone?' I hadn't even heard of it ... Professor Tingley was greatly disappointed, and I did not understand why until a few weeks later he called the student body together to explain and illustrate the telephone by a homemade instrument. 'You'll talk to your homes from these rooms one day,' he told us. 'New York will talk to Boston.' He didn't suggest Chicago. 'Dreamer,' the boys said.

Ida Tarbel: *All in the Day's Work: An Autobiography* (1939)

After the huge success of the telegraph, it seemed inevitable that efforts would be made to carry more than a series of dots and dashes down a wire. Why not voices too? Following the triumphant completion of the initial transatlantic telegraph cables in the 1860s, and combined with a growing understanding of the physics of electricity and magnetism, new possibilities for using the electron to communicate flourished. Progress, though, was not as straightforward as this summary makes it sound.

Making room for more messages

As telegraphy took off in a big way, the biggest problem faced was capacity on the wires. There was so much demand that the most popular lines were getting far more messages than their operators could send. It was expensive and cumbersome to run more and more cables between telegraph stations, leading a whole range of inventors to search for ways to send multiple signals down the same cable simultaneously.

Perhaps the most sophisticated attempt to do this without radically changing the technology came from Thomas Edison in 1869, an invention that he sold to the Atlantic and Pacific Telegraph Company for the considerable sum of $30,000* in 1874. This was a so-called quadruplex telegraph, which could send two signals each way down a single cable at the same time – as the name suggests, this quadrupled the traffic a single wire could handle.

The ability to send in both directions had been made practical a little earlier by another American inventor, Joseph Stearns. This was done by sending a cancelling signal to the local, receiving relay so that only the remote station received a message. But Edison added the ability to have two separate messages in each direction by switching the positive and negative polarity for the second message.

Edison's approach did work, but it was frequently glitchy and, more importantly, could not be extended beyond two messages at a time in one direction. The field was open to find a different way to offer multiple signals (known as multiplexing) – and the guiding principle was to make use of frequencies. This could be approached in two ways. As we have seen, although the early telegraph was intended to use marks produced by a relay, pulling a marker down onto a paper roll while an electric current was applied, the operators found it easier to rely on the clicking sounds these relays made. What was needed was either to have relays sensitive to specific frequencies, or to produce an audible tone, allowing a deep tone, say, to be distinguished from a high-pitched one.

The purely auditory approach, while theoretically usable, was limited to an operator's ability to mentally tune out different frequencies – with the output of several operators producing different frequency beeps simultaneously, the result was likely to be more like a piece of avant garde music than a distinguishable set of messages.

* That's around $880,000 in 2025 with inflation, or around $9.6 million in terms of salaries.

The practical use of what was known as acoustic or harmonic telegraphy was largely confined to having detectors tuned to a particular frequency. Edison was among those attempting a solution again, experimenting with devices that made use of tuning forks to select out a signal, but the real battle to produce a workable mechanism, which ironically would render the telegraph largely redundant, was between Scottish-born American inventor Alexander Graham Bell and the all-American inventor Elisha Gray.

Bell versus Gray

Gray was co-founder of the company that would become Western Electric and had been working on telegraphy equipment since 1865. Bell, by contrast, had been a teacher of deaf-mute pupils and possessed no engineering background, though his focus on the deaf, driven by a family history of speech researchers and having a deaf mother, had led to an interest in the science of acoustics. He came to telegraphy through an emotional attachment.

One of Bell's pupils, Mabel Hubbard, was the daughter of high-flying lawyer Gardiner Greene Hubbard, who had a financial interest in telegraphy and was concerned by the market dominance of Western Union. As Bell and Mabel became closer, Bell visited the Hubbard home and was able to make use of his acoustic expertise to suggest that a telegraph could be multiplexed by using harmonics, much as a single string on an instrument like a violin could produce different notes. With some funding from Hubbard, Bell took on an 'assistant'. This was Thomas Watson, who had the electrical engineering expertise Bell lacked, so arguably had a more significant role in the partnership than his job title suggested.

Despite Watson's help, though, progress was slow, not helped by the pressure Bell faced when Hubbard effectively refused Bell's request for his daughter's hand in marriage unless Bell could first come up with a commercially viable harmonic telegraph. It was around this time – late 1875 – that Bell first seems to have made the mental leap from the concept of sending different frequencies down a wire to the idea that, were this possible, he could transmit audio signals. This was, after all, just a collection of frequencies, and speech was a much more natural way of communicating than dots and dashes. However, this was seen by Hubbard as an unnecessary frippery, moving focus away from the commercial opportunity.

Meanwhile, Elisha Gray had been developing a similar harmonic telegraph which used metal strips vibrating at a range of frequencies to produce electrical signals that would start similarly tuned strips vibrating at the other end of the wire, enabling multiple signals to be sent simultaneously. Gray got an 1875 patent for an 'Electric Telegraph for Transmitting Musical Tones'. From this, he clearly moved on to think about transmitting sounds in a similar way.

It was at this point that proceedings became murky. According to one version of the story, in February 1876, Gray submitted a patent application* entitled 'Instruments for Transmitting and Receiving Vocal Sounds Telegraphically'. Note that this was explicitly a telephone – the accompanying diagram even shows a moustachioed man, apparently speaking into the apparatus. With his network of contacts, Hubbard heard about this rival development and allegedly bribed an official to accept a vague patent application from Bell titled 'Several improvements on the telegraph', which was purely focused on harmonic multiplexing, underhandedly getting it registered as received before Gray's application.

The story then goes that, at Hubbard's instance, Bell visited the patent office in Washington and (again with an allegation of bribery) modified his application post-submission to include some details that would make his invention more appropriate for sound transmission, based on the contents of Gray's patent caveat. Less than two weeks later – a remarkably short period of time unless there were indeed financial incentives involved – Bell's patent, with these modifications, was approved.

Exactly what really happened here, though, is disputed. The patent examiner was apparently an alcoholic who was in debt to Bell's lawyer. He swore an affidavit that he had accepted payment to say that Bell's application came first and to pass on details of Gray's mechanism. But he later seems to have denied this – and the affidavit was apparently

* Technically, Gray submitted a 'patent caveat' – effectively a holding statement before a full patent was able to be submitted.

drawn up by lawyers from one of Bell and Hubbard's competitors. Despite some similarities in the liquid transducer (more on this in a moment), the detail of the patents was significantly different – and Bell's was approved because Gray withdrew his patent caveat, not because of dubious action on the part of Hubbard.

Transforming sound to electricity

To be able to transmit varying frequencies, Gray's patent application had described a special type of electrical resistor that used a liquid mechanism. This would act as a transducer – a device for changing one type of energy (in this case sound waves) into another (electrical currents). According to the anti-Bell faction, Bell and Watson hurriedly began attempts to duplicate this approach. However, Bell had already filed a patent the year before for a different device that made use of liquid transducers, making the suggestion that he stole the entire concept from Gray unlikely.

The liquid transducer that Bell had worked on in 1872, based on some work by German scientist Hermann von Helmholtz, used the vibrations of an electrically stimulated tuning fork to dip a wire in and out of a liquid, most often a dilute acid, making and breaking a circuit to produce an electrical signal that varied at that particular frequency. This signal was fed down a wire to a second tuning fork which reproduced the sound. By 1873, the water component had been replaced by mercury and Bell was beginning to replace the tuning fork with a vibrating metal strip, where the vibration was initiated using sound waves, speaking close to the metal strip.

As experiments went on, a flexible gutta-percha mouthpiece was used instead, which repeatedly dipped a wire into the liquid as the rubbery mouthpiece was vibrated by sound waves. During his experiments on transmitting sound using a liquid transducer, Bell called out to Watson, 'Mr Watson, please come here, I need you!' This message was said to have been carried unexpectedly to another room by the nascent telephone, making it the first recorded call. But like the rest of the Bell mythology, there may be exaggeration at play with the apparent dramatic accident being instead an intentional test.

Before his first public demonstration at the International Centennial Exposition, held in Philadelphia in June 1876, Bell was able to switch to a magnet-based mechanism to convert sound into electrical signals. It has been said that this was no better than the liquid approach, but it meant that he was no longer dependent on a stolen technology. Again, it depends which version of the story you prefer; in reality, the magnetic approach was both safer and far more versatile than using the liquid transducer.

Hubbard was now pressing forward from the business end and was able to purchase a more effective mechanism for conversion of sound to electricity. This was the invention of German-born American inventor Emile Berliner, whose big claim to fame would be the invention of the gramophone.* For Bell, exciting though the technological breakthrough might have been, the most important outcome was that Hubbard now allowed his marriage to Mabel to go ahead – he was happy to leave Hubbard to deal with the expansion of the business.

In both the telegraph and the telephone, the electron was the carrier of information, conducting a signal down a wire. But other developments were beginning to put into question the need for that wire in the first place. Once again moving electrons would be involved, but in this case the interaction would not be with a Morse key or sound, but with electromagnetic waves, the phenomenon responsible for light.

* Emile Berliner invented the gramophone, which used discs, as opposed to Edison's earlier invention of the phonograph, which played the less practical cylinders. I once had a lengthy dispute in a pub quiz where my answer to the question 'Who invented the gramophone?' was judged incorrect because I didn't say Edison.

14 GOING WIRELESS

After a few preliminary experiments with Hertzian waves I became very soon convinced, that if these waves or similar waves could be reliably transmitted and received over considerable distances a new system of communication would become available possessing enormous advantages over flashlights and optical methods, which are so much dependent for their success on the clearness of the atmosphere.
Guglielmo Marconi, Nobel Lecture: *Wireless telegraphic communication* (1909)

To move our electron-inspired communications beyond wires, we first need to take a step back to 1861, when Scottish physicist James Clerk Maxwell left London for his annual summer break at Glenlair, his family estate in Scotland. At the time, Maxwell was wrestling with a model he had been developing for some time to represent the interaction between electricity and magnetism. It involved a combination of rotating hexagonal cells, surrounded by small, ball-bearing like spheres, collectively representing the aether.

Meet the aether

It's worth taking a moment to introduce the aether,* as it will feature significantly along the way as the electron was beginning to emerge as

* The aether is sometimes spelled 'ether', a spelling that gradually took over in the later part of the nineteenth century and was nearly universal by the

a character worthy of a biography. The usual rule is that waves require a medium to travel through – think of waves in the sea, for example. If you took away the medium (in this case, the water) there could not be a wave. With sound waves the medium is the air – again, take away the air and sound can't travel. In space, as the tagline for the film *Alien* had it, no one can hear you scream. But light somehow manages to traverse the empty vacuum of space.

To account for this, a substance known as the aether (or to give it its full title, the luminiferous aether) was dreamed up. This was an invisible, intangible substance that filled all of space, which provided the medium through which the light waves could travel. Though insubstantial, the aether would also have to be incredibly rigid, as waves lose energy and die away in floppy mediums – yet also it would have to flex to enable light to travel vast distances through space, a property more usually expected to be limited to particles of matter.

The existence of the aether should have been detectable, which scientists Albert Michelson and Edward Morley, working in Cleveland, Ohio, set out to do in 1887. They used a massive piece of technology for this. First came a solid brick base, cemented into place on the laboratory floor. On top of this was a circular metal trough, filled with mercury. In the trough floated a circular construction of wood, nearly touching its sides, and finally on top of the wooden circle was a slab of stone over a metre across. The whole device was designed to protect the slab from any vibration and to keep it moving steadily once it started. Such was the lack of friction in the carefully constructed mercury trough that once the slab was rotating at around a turn every six minutes it kept moving for hours.

On top of the apparatus, a series of mirrors enabled a beam of light to pass backwards and forwards across the slab. The beam was split into two, with half tracing a path in one direction, the other half travelling at right angles, before the two rays were brought back together. These two beams of light then interfered with each other, producing

---

start of the twentieth century. But I prefer aether, both because it makes it sound suitably antiquated to fit with the period, and also because it avoids confusion with the chemical substance used in early anaesthetics more properly called diethyl ether with the formula $(CH_3CH_2)_2O$.

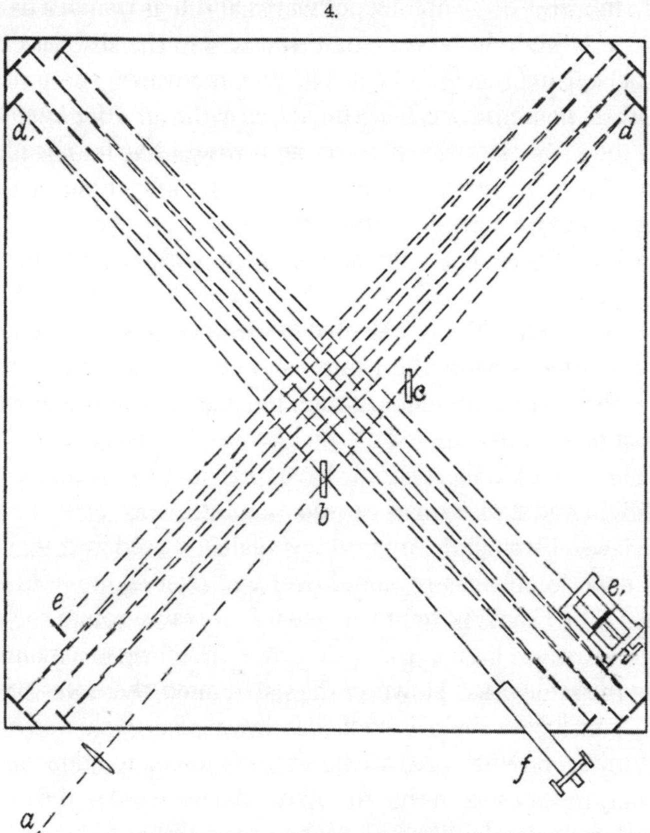

Michelson–Morley experiment.

Albert Abraham Michelson (1852–1931) with Edward Morley (1838–1923),

Public domain, via Wikimedia Commons

a pattern of fringes that were viewed through a small microscope mounted on the block.

Interference is a common way that waves interact. If you imagine, for example, dropping two stones into a still pond an arm's length apart, the ripples caused by each stone will head outwards. Soon, these waves originating from each stone's entry into the water will start to meet. If at a point of meeting both waves are moving up or down in the same direction then they will reinforce each other, creating a larger wave. If, though, they are travelling up or down in opposing directions they will cancel each other out. This produces a series of visible 'fringes'. In water these are deeper waves and regions with hardly any movement. With light waves, these appear as light and dark bands.

Michelson and Morley's heavyweight experiment was supposed to demonstrate something called the aether wind. As the Earth moved through the aether it was expected that it would give light travelling in the same direction as the movement a slightly longer path to traverse before it caught up with a point on the planet's surface. The Cleveland device was designed as an optical racetrack to detect this effect.

The same beam of light, split, was sent off on two paths, each exactly the same length but oriented 90 degrees away from each other. At any one time, one of the beams should be oriented more in the direction the Earth is moving through the aether than the other, so the timings should have changed as the block slowly rotated. This meant that one beam would increasingly have further to travel than the other, and the light and dark fringes caused by interference between the two beams, viewed through the microscope against a fixed grid, would shift.

The experiment was repeated over and over again – yet nothing happened. The beams of light threaded their way through the maze of mirrors and arrived back at the microscope. The fringes remained stubbornly in the same place. However the slab rotated, there was no change. The only conclusion the pair could reach was that there was no aether wind. With deep reluctance, Michelson was forced to admit that quite accidentally he had shown that the aether did not exist in the first place.

It took considerable time, though, for the aether to be entirely dismissed. Maxwell never gave up on its existence, even though his work eventually made it theoretically unnecessary. A younger British physicist, Oliver Lodge, who would make a significant development to Maxwell's equations, was still convinced enough of the aether's existence in 1909 that he wrote a book called *The Ether of Space*, where he told the world that the aether was 'not only uniformly present and all-pervading, but also massive and substantial beyond conception ... by far the most substantial thing – perhaps the only substantial thing – in the universe'.

Maxwell's model interaction

Let's get back to Maxwell's model of the interaction between electricity and magnetism, represented by a combination of rotating hexagonal cells, surrounded by small, ball-bearing-like spheres. This was designed to reflect the way that an electrical current starting to flow through a wire would create a magnetic field, which then induced a current in another wire. In Maxwell's model, rows of hexagonal cells rotated against each other under the influence of the ball bearings. Unfortunately, though this mechanical analogy worked well for induction, it failed to reproduce other interactions between electricity and magnetism.

Bearing in mind there was no suggestion that there really existed a set of invisible rotating cells and ball bearings – this was merely a mental mechanical representation – it was entirely possible that Maxwell was trying to get too much from his physical analogy. Whether it was being able to spend more time with his wife Katherine, who always proved an important sounding board for his ideas, or getting away from busy London to the relaxed Scottish countryside where he would always feel more at home, Maxwell came up with a variant of his model where the hexagonal cells were not rigid, but instead were flexible.

If this were the case, he realised that a twisted cell would shrink top to bottom – corresponding exactly to the phenomenon of electrostatic attraction, if he imagined the aether between a pair of electrically charged objects contracting as the cells were twisted. Similarly, this enabled him to model a condenser (now known as a capacitor), the successor to the Leyden jar as a way of storing electrical charge. This was impressive in itself – but inspired him to consider something else, something even more significant to the future of physics.

An elastic medium is what's necessary to carry a wave. As we have seen, a wave involves regular movement of matter in a medium – water, for example, for waves in the sea – and that movement requires the substance involved not to be entirely rigid, or that waving motion could not occur. In Maxwell's model, the cells were the constituents of magnetism, while the ball bearings made up electricity. If we imagine a twitch in a row of ball bearings, this would move the adjacent cell, twisting it. The cell would then spring back, moving more ball bearings and so on. An electric wave would be generating a magnetic wave, which generated an electric wave and so on.

It's not clear whether or not Maxwell knew of an impromptu lecture that his hero Michael Faraday had made at the Royal Institution when Maxwell was just fifteen. Faraday had to speak instead of a missing lecturer and ran out of material on the planned topic (a kind of electric clock). To fill in, Faraday speculated that light was a form of vibration in the electrical and magnetic fields. In fact, in his speculation he had gone further than Maxwell ever would in realising that such an interacting wave would not need the medium of the aether. As Faraday had put it, '[The theory] endeavours to dismiss the aether but not the vibrations.'

Theorising at the speed of light

Either way, it seemed to Maxwell that it was entirely possible that the wave predicted by his model could be light. And he had a way to check this out. If he could work out the elasticity and density of the aether predicted by his model, he would be able to calculate the speed at which these waves would travel. These values could be approximated to by what was known about the units of electrical and magnetic charge and the strength of electrical and magnetic forces. He worked out that, in a vacuum, this speed would be 193,088 miles (310,700 kilometres) per second. If this were the same as the speed of light, it would surely show that light was an electromagnetic wave.

Over the years, many attempts had been made to measure the speed of light. Galileo, for instance, tried to time light by flashing a signal to a distant observer, waiting until he saw a returning flash and timing the interval in between. Unfortunately, he discovered the same timing when his assistant was much closer – what he was measuring was human reaction time, while light was far faster.

By 1861, though, there was a reasonable idea of the speed of light, calculated by French physicist Armand Fizeau by sending light through a fast-rotating toothed wheel, producing short flashes, down a 9-kilometre track then back through the wheel, making it possible to deduce the speed of light by the amount the wheel had turned before the flash returned. Unfortunately, Maxwell did not have a book or paper at Glenlair that would have told him Fizeau's figure.

It was only when Maxwell returned to King's College in London for the October start of term that he was able to look up Fizeau's result, giving light's speed as 195,647 miles (314,850 kilometres) per second.

Other experiments around that time estimated the speed at between 192,000 and 193,118 miles (308,990 and 310,790 kilometres) per second. Bearing in mind there was some estimation in his calculation, being within 1.5 per cent of the expected value was a remarkable confirmation of his theory.

Maxwell died tragically young, aged just 48, never having the chance to prove the existence of electromagnetic waves. However, by 1873 he had crystalised his theory of electromagnetism into a mathematical form that did away with his mechanical models. It would take more than a decade more before anyone could support his theory with experiment. This was down to Heinrich Hertz, the professor of physics at Karlsruhe in Germany, an expert on Maxwell's theory who had helped simplify Maxwell's equations of electromagnetism. In 1879, Hertz had been pointed in the direction of electromagnetic waves by his doctoral supervisor Hermann von Helmholtz (whom we last met devising liquid transducers with tuning forks).

Generating Hertzian waves

In 1887, Hertz wrote up a demonstration that would not have looked out of place in Frankenstein's lab. He generated a spark by putting a high voltage across a gap in a brass rod. At the other side of room he had a second rod, again with a (narrower) gap. As he had hoped, a faint spark began to jump across the second gap. The initial spark was generating an electromagnetic wave, which crossed the lab, interacted with the distant split rod and starting an electrical current in that. Although Hertz had no interest in using this practically,* these waves, initially known as Hertzian waves, were the first intentional generation of radio waves.

We've not mentioned the electron for a while: it was present and hard at work here, though it was yet to be explicitly discovered. When an electrically charged particle like an electron is accelerated, it gives

* Hertz is said to have commented of his radio wave discovery, 'It's of no use whatsoever,' and when asked about what might be done with it is said to have replied enthusiastically, 'Nothing, I guess.'

off energy in the form of an electromagnetic wave (or a stream of photons if we take a quantum viewpoint). In a radio aerial, electrons are accelerated up and down the metal in its structure. As they do so, radio waves stream off. When those waves hit the receiver aerial, their energy is transferred to electrons in the metal, starting a new, oscillating electrical current.

Though known as Hertzian waves for some time in the scientific world, the first practical use (which we'll come across in a moment) was referred to as wireless telegraphy, leading to an early term for a radio receiver of a 'wireless', a term that lingered in the UK until the 1960s.* The waves themselves were renamed radio waves, a term taken from the Latin word radius, which as well as being the spoke of a wheel (hence the radius of a circle) also referred to a beam of light. The usage began in the 1890s, becoming commonplace by the mid-1900s. Over time, 'radio' would replace 'wireless' as the name for the receiver itself.

There were a number of attempts to provide wireless electrical communication as the nineteenth century drew to a close, mostly relying on the short-range capabilities of electrical induction. Although Nikola Tesla (whom we will return to again in much more detail in the next chapter) was largely dismissive of radio as a communications medium, he constructed the first radio-controlled vehicle – a boat – in 1898, calling his remote control 'teleautomation'. But despite grand visions for its capabilities, he was distracted by attempts to provide long-range communication via a form of purely electrical waves, leaving the radio field clear for the young Italian Guglielmo Marconi.

Tesla, incidentally, always felt slighted by Marconi's achievements, once commenting that his rival's efforts were 'just suitable for the amusement of small boys who are beginning their electrical experience'. He would several times, baselessly, challenge Marconi's patents.

* For me, the death of 'wireless' coincided with the introduction of solid-state electronics. My parents' valve-based wireless was replaced by a transistor radio.

The making of Marconi

By contrast, Marconi always had a better business focus than Tesla, taking the slow, steady steps necessary to achieve his goals. He was born in Bologna in 1874 to a wealthy family. Though always a proud Italian, his family had strong ties to the English-speaking world – his mother was Irish (from the Jameson whiskey dynasty) and his father took British citizenship in 1877. As a result, it was in the UK that Marconi would start what would become one of the country's biggest businesses. There would remain a Marconi company in England until 2006, when its telecommunications arm collapsed, seven years after Marconi Electronic Systems merged with British Aerospace to form BAE Systems.

Guglielmo Marconi had first achieved a small success by ringing a bell from across his bedroom in 1894 (allegedly with the assistance of the butler, Mignani), aged just twenty. Step by step, he built up his capabilities. Within a year he had realised the benefits of grounding his apparatus and put up a taller aerial, managing to send a signal over a distance of 3 kilometres (2 miles). Frustrated with a lack of funding opportunities in Italy, in 1896 he moved (with his mother) to England.

Here, supported by the chief electrical engineer of the General Post Office, William Preece, who was already an enthusiast for telegraphy, Marconi was able to make demonstrations with increasing range and scope. In 1897, when he also set up the Wireless Telegraph and Signal Company (later renamed the Marconi Company), he sent the first message to a ship. In 1899, his radio waves had crossed the English Channel to France, making the first international wireless communication. By 1901 (or possibly 1902, as there is some doubt about the reception in the earliest experiments), he had achieved the big breakthrough. Less than 40 years after the first successful transatlantic cable was laid, he had transmitted a wireless signal from Poldhu in Cornwall to Signal Hill in Newfoundland.

Just as conventional telegraphy had rapidly been overtaken by telephony, so the initial Morse code wireless telegraphy would be replaced by audio forms of radio for speech and music,* later joined

* Morse code radio signals continue to be sent in the present day by some specialists, including radio amateurs.

by a visual component in TV, also carried on the same electron-driven concept of radio. It would be around twenty years before radio broadcasts began to take over from telegraphy, a development in part due to another electron achievement we will first meet in chapter 23, the first generation of electronics.

In 1909, Marconi's achievements were cemented by the award of the Nobel Prize in Physics. This, perhaps more than anything else, hugely irritated the American-based Serbian inventor Nikola Tesla. He was convinced that Marconi's work was little more than a distraction from an electron-based technology that not only could communicate to anywhere in the world, but also would make it possible to transmit unlimited amounts of energy to remote locations.

15 THE WARDENCLYFFE FOLLY

The scientists from Franklin to Morse were clear thinkers and did not produce erroneous theories. The scientists of today think deeply instead of clearly. One must be sane to think clearly, but one can think deeply and be quite insane.

Nikola Tesla in *Modern Mechanics and Inventions* (1934)

There is no more divisive figure in the history of electricity than Nikola Tesla. A few minutes' search on the internet will uncover wide-ranging claims that he was an underrated genius, achieving remarkable things far ahead of his time. You might discover that he 'invented' X-rays before Röntgen, received a radio signal from Mars, wirelessly transmitted signals and energy worldwide, could send energy faster than light and designed a motor that ran on cosmic rays. There is no evidence for any of these assertions. The myth of Tesla remains strong, though.* More importantly from the factual viewpoint, though, Tesla was also an innovative electrical engineer who made major contributions to the field. We need to unpick the fantasies from the reality, to discover the true outcome of Tesla's love affair with the electron.

---

* There is an excellent portrayal of Tesla's combination of showmanship, technology and outlandish claims in the film *The Prestige*, based on the Christopher Priest book of the same name. David Bowie's Tesla creates a teleportation device central to the plot, but a scene where Tesla remotely lights up a field full of bulbs is a perfect illustration of the real Tesla's approach.

As we saw in chapter 12, Tesla was born in Smiljan in 1856, partly completing a course in engineering before moving to Paris to work for Edison. This led to his emigration to New York, continuing to work for the company. Soon after, he set up his own business which lasted less than a year – but this was the start of an entrepreneurial life, often working with large companies and leading financiers, but always setting his own agenda.

Our first venture into the debate over AC versus DC was primarily from Edison's DC viewpoint, but for Tesla, the advantages of AC were never to be doubted, and in 1887 he produced his greatest invention in the first truly effective AC motor. Unlike transmission, DC has the natural advantage here. The first electrical motors were all DC, but Tesla produced an efficient AC motor making use of induction to drive a rotor. This would lead to a significant ten-year patent contract with Westinghouse, which would later be followed by a swathe of lucrative patents.

Transmitting electricity wirelessly

Tesla did experiment with X-rays soon after their discovery, and as we have seen would also demonstrate what was probably the first radio-controlled boat. But his other big invention was the Tesla coil, a way of generating high-frequency, high-voltage AC which enabled the production of huge electrical sparks and was able to induce a sufficiently strong current to light up bulbs remotely. This was particularly useful as the creative electrical engineer was transformed to a full-on showman by exposure to US culture.

At the time, radio was becoming a significant phenomenon – making use of the interaction between moving electrons and electromagnetic waves – but despite his radio-controlled boat, this was where Tesla began to deviate from the scientific understanding of his day (or, for that matter, of the present). As the opening quote of this chapter shows, Tesla was highly doubtful of twentieth-century physics – he was very much fixed in the Victorian mindset as far as science goes.*

* It's also interesting that in the Tesla quote at the opening of the chapter that his named 'scientists' are both men with little or no scientific training.

Tesla refused, for example, to accept the validity of relativity or quantum theory. He was scathing of Einstein, commenting that his relativity work 'is a magnificent mathematical garb which fascinates, dazzles and makes people blind to the underlying errors. The theory is like a beggar clothed in purple whom ignorant people take for a king.' And in a similar fashion, despite his engineering genius, Tesla would demonstrate a total lack of understanding of the underlying nature of electricity and magnetism.

This deviation between accepted theory and Tesla's ideas came to a head with his attempt to go one better than radio. Tesla doubted that radio could work at long distances because it transmits signals in a straight line, so should have been thwarted by the curvature of the Earth.* Rather than make use of electromagnetic waves to communicate remotely, Tesla envisaged that he would be able to transmit raw electrical power all around the world, much as he did to nearby light bulbs. He thought that this would be both more reliable than radio for communication and make it possible to provide energy wirelessly at any distance, doing away with expensive power cables and pylons.

In his first experiments, he claimed that he had transmitted electrical energy across tens of miles, though suspiciously no one was ever allowed to see the process in action, not even his assistants. But his big throw of the dice on wireless power transmission was Wardenclyffe. This was a site on Long Island where he constructed a distinctive electrical transmitter with a high tower topped by a bulbous cupola, funded by a combination of $33,000 of his own money, a $10,000 mortgage and a $150,000 investment by financier J.P. Morgan in return for 51 per cent of the patent rights.

Waving the Earth

Tesla's idea was to generate electrical waves that would be tuned to the resonant frequency of the Earth. This resonance, he believed, would

---

* In practice, radio was more flexible than Tesla thought, due to reflections off the charged ionosphere in the upper atmosphere, and some diffraction effects.

amplify the waves to an extent that would enable them to reach anywhere in the world. The waves would pass through the ground before completing a return circuit through the air, making use of a conducting layer in the upper atmosphere. In physical terms this made no sense – he was thinking in terms of a wired electrical setup, where a complete circuit is required for the electrons to flow. But electrical and electromagnetic waves are a totally different proposition.

Wardenclyffe, in the town of Shoreham, had been a resort site that failed, but Tesla bought the site and built a laboratory building that would be dominated by a 57-metre (186-foot) high tower topped with a strange-looking steel or copper hemisphere 21 metres (68 feet) across (reports differ on its composition). The final look was distinctly

The Wardenclyffe tower.

Public domain, via Wikimedia Commons

phallic. Because the idea was to feed into the earth below, there were massive works beneath the surface, with metal pipes radiating out at least a hundred feet, intended to enable the resonant waves to be produced. Tesla described their function as 'having a grip on the earth, so the whole of this globe can quiver'. The transmitter was constructed between 1901 and 1902, but never fully financed. As a result, Tesla had to finish operations in 1906. The tower was demolished in 1917 and he lost the site in 1922, though the laboratory building still stands and was later used for industrial purposes.

Tesla was, without doubt, a strange man. He was obsessed with pigeons, which he looked after in his hotel rooms. Humans he was less sure about, insisting that hotel servants stayed at least a metre (3 feet) away from him. He was only happy to eat if he had calculated the volume of his cup, soup bowl and food consumed – and if a fly landed on his table, the whole thing had to be reset. His scientific understanding was limited. But his brilliant electrical engineering abilities enabled him to make full use of the impact of the electron's back-and-forth dance that would result in induction.

By contrast, our next electrical pioneer, William Crookes, was not only a very capable physicist, but also the first to begin to uncover an aspect of what electricity truly was, with the first hint of the existence of the electron itself. Yet for some observers, his experimental expertise was tainted by his enthusiasm for spiritualism.

THE MYSTERIOUS DR CROOKES

16

> *Crookes received the high British distinction of the Order of Merit in 1910, and in 1913 he was elected President of the Royal Society. Crookes was esteemed not only in scientific circles but by the general public, and the tag* Ubi Crookes, ibi lux *[Where Crookes is, there is light] reflected popular feeling.*
>
> Eric Holmyard – *British Scientists* (1951)

William Crookes began his scientific life as a chemist. Born in 1832 in London, by the age of sixteen he had begun studies at the Royal College of Chemistry (now part of Imperial College, London). This led to an interest in the still-young field of photography: by 1857 he was working as a photographic chemist. Experimenting from his home laboratory, he discovered and named the element thalium in 1861 (co-discovered in the same year by French chemist Claude Lamy) and was elected a Fellow of the Royal Society in 1863.

Tubes and spirits

Crookes' name has remained attached most clearly, though, to the device that links him to electrons and bears his name: the Crookes tube. But before encountering this in more detail, we need to bring in another important aspect of his life: spiritualism. His interest in the peculiarly Victorian phenomenon was perhaps linked to the death of his younger brother Philip in 1867, which was around the time Crookes began to explore the field.

He was not alone among scientists of the period in having an interest in paranormal phenomena – for example, the younger physicist Oliver Lodge, who helped standardise Maxwell's electromagnetic equations, was fascinated by the possibilities of both telepathy and communication with the dead. It was a topic to which many big names of the late Victorian and Edwardian period succumbed, notably also Arthur Conan Doyle, the author of the Sherlock Holmes stories, who was sufficiently entranced that he publicly supported the now infamous Cottingley fairy photographs faked by young cousins Frances Griffiths and Elsie Wright.

Crookes first investigated a then-famous medium called Daniel Dunglas Home, writing up his experiences in the *Quarterly Journal of Science* and attempting to apply a scientific rigour to the claims of mediums – but sadly, like many subsequent scientists, he proved far less effective at spotting tricksters than stage magicians like Harry Houdini. Crookes went on to support the work of a young female medium named Florence Cook, who claimed to materialise the spirit of a girl named Katie King. More sanguine observers noted that the spirit of Ms King looked remarkably like Cook, but Crookes was convinced of her abilities.

It has been suggested that Crookes had fallen in love with Cook and so was easily manipulated into seeing what she wanted him to see. Alice Vernon, in her book *Ghosted*, suggests, 'Crookes defended Florence almost to the point of ruining his own career. He was relentlessly hounded by critics even well into the twentieth century when it was revealed that Crookes' opinion of Florence may have been helped by a few secret intimate exchanges.'

According to the physicist Frank Close, 'Convinced that during seances [Crookes] had seen "luminous green clouds" and that the lights in his vacuum tube were the same as these phantoms, he announced in 1874 that he had produced ectoplasm.'* In later

* Ectoplasm was a supposed substance of spiritual energy produced by some spiritualist mediums, usually found to be cheesecloth or similar materials when debunked. Some such 'ectoplasm' was treated to glow green in the dark. Crookes' belief is described in *Destroyer of Worlds*, Frank Close (2025), where it is attributed to Crookes without a source.

years, and in a more scientific manner, Crookes certainly wondered about a connection between the phenomena exhibited in Crookes tubes and telepathy. In an address to the Society for Psychical Research (of which he was then president) in 1897, Crookes commented:

> These rays, as generated in the vacuum tube, are not homogeneous, but consist of bundles of different wave-lengths, analogous to what would be differences of colour could we see them as light. Some pass easily through flesh, but are partially arrested by bone, while others pass with almost equal facility through bone and flesh.
>
> It seems to me that in these rays we may have a possible mode of transmitting intelligence which, with a few reasonable postulates, may supply a key to much that is obscure in psychical research. Let it be assumed that these rays, or rays even of higher frequency, can pass into the brain and act on some nervous centre there. Let it be conceived that the brain contains a centre which uses these rays as the vocal chords use sound vibrations (both being under the command of intelligence), and sends them out, with the velocity of light, to impinge on the receiving ganglion of another brain. In this way some, at least, of the phenomena of telepathy, and the transmission of intelligence from one sensitive to another through long distances, seem to come into the domain of law and can be grasped ... By such an hypothesis no physical laws are violated; neither is it necessary to invoke what is commonly called the supernatural.

The inspiration for this thinking, the Crookes tube, was a sealed glass vessel, often shaped like a partly inflated long balloon. The air inside the tube was mostly pumped out leaving only a low-pressure gas. At one (usually narrower) end a negative electrode (cathode) entered the tube, while near the other end was a positive electrode (anode). It was not uncommon to have a piece of metal partway down the tube, often shaped like a Maltese cross. It has been suggested that this distinctive shape, so strongly associated with the Crookes tube, was selected because the Latin word for cross, crux, sounds similar to the scientist's name.

Detecting cathode rays

For decades, the sometimes beautiful electrical discharges in tubes containing low-pressure gases had been studied with limited understanding of what was involved. In historical times, this was nothing new. Think, for example, of the early explorations of electricity and magnetism, which were seen as wonders, worth exploring for their sheer fascination. Michael Faraday had discovered what he described as a 'glow discharge' in a partially evacuated tube in 1838, while by 1857 the German physicist Heinrich Geissler, who had a much better vacuum pump, was able to produce the kind of bright radiance familiar from neon lights.

Crookes had an even better vacuum pump, and he made use of higher voltages. In his tube there was no glow from the remnants of gas. Instead, it was the glass itself at the far end of the tube that glowed a bright green. Because of this, Crookes assumed there were 'cathode rays' emanating from the cathode and flowing down the tube to impact the glass. This idea was reinforced by putting fluorescing materials or metal shapes in the tube, which would respectively glow and cast a shadow when the rays were blocked. So strongly embedded was the term that even when it was entirely clear what was involved, 'cathode rays' was still used to describe them, hence the tube in traditional TVs and computer monitors being known as cathode ray tubes (CRTs).

Note in Crookes' quote above, he referred to rays where 'Some pass easily through flesh, but are partially arrested by bone' – a clear reference to X-rays. These had been discovered two years before he gave his talk by German physicist Wilhelm Röntgen, who was using a Crookes tube in his experiments. X-rays are electromagnetic radiation – a form of light – in this case produced by high-energy electrons smashing into metal or the glass walls of the tube where they pushed atomic electrons to higher level orbitals, giving off X-rays as the electrons fell back.

The concept of 'rays' now feels distinctly old fashioned. By the start of the twentieth century at least five different types of ray had been identified. Apart from X-rays, the burgeoning concept of radioactivity was responsible for three types of ray: alpha, beta and gamma, all operating alongside the cathode ray. Over time, true rays would become limited to electromagnetic radiation – in effect, a range of energies of light that ran from radio, through the visible, reaching at the high-end gamma rays. The rest would be pinned down as streams of particles, with two out of three (cathode and beta) proving to be electrons.

Of course, this wasn't known at the time, and there was considerable dispute over exactly what cathode rays were. Crookes himself believed that the electrical voltage was giving a charge to remaining atoms in the tube, which were then pulled down in the direction of the anode – what he would describe as radiant matter. Others were more inclined to put the rays down as vibrations in the aether.

The paddle wheel error

Crookes thought at one point that he had dreamed up a way of demonstrating the material nature of cathode rays. In 1879, he placed a paddle wheel partway down the tube so it would be hit by oncoming cathode rays: as a result, the wheel started to turn. Crookes triumphantly announced this was evidence of the existence of particles rather than waves in the aether. Unfortunately, he was making a mistake that would be compounded by a toy still available today, known as a Crookes radiometer.

The radiometer is an evacuated glass tube shaped like a light bulb, which contains a vaned paddle wheel, with each vane painted

Crookes' paddle wheel tube.

William Crookes, Public domain, via Wikimedia Commons

white on one side and black on the other. When the radiometer is placed in sunlight, the paddle wheel begins to vigorously rotate. This is often explained as being a demonstration of the way that photons of light, while not having any mass, do have momentum and so can push the wheel around as the photons bounce off the white side, meaning the paddle feels an excess pressure on that side.

It's true that light can produce pressure – in space, 'light sails' can be driven by sunlight. But unfortunately, the pressure isn't sufficiently strong to move the paddles in a radiometer. Worse still, if you take a look at a Crookes radiometer,* there is something painfully wrong with this theory. The paddles rotate in the opposite direction to that expected. Rather than excess momentum being given to the white side, it is the black side absorbing energy and heating up that causes the movement. The heat is transferred to nearby gas molecules, which get extra energy, so bump more strongly against the black side, applying pressure.

Although Crookes' paddle wheel in a cathode ray tube did not have the black/white distinction and so rotated the 'correct' way, it was, again, a result of the rays' impact heating up the paddles, which transferred energy to nearby gas molecules, impacts from which started the paddle wheel rotating. As such effects could also be created by light, at the time only thought of as waves, the experiment gave no evidence towards the particulate nature of cathode rays.

By 1893, when English physicist J.J. Thomson began to write about such discharges and cathode rays, he was making a conscious attempt to distinguish between the two theories and to move on from the visual drama of Crookes tubes to discover the underlying physics. In *Notes and Recent Researches in Electricity and Magnetism* Thomson wrote, 'The phenomena attending the electric discharge through gases are so beautiful and varied that they have attracted the attention of numerous observers. The attention given to those phenomena is not, however, due so much to the beauty of the experiments, as to the widespread conviction that there is perhaps no other branch of physics which

---

* You can see a Crookes radiometer in action at https://universeinsideyou.com/experiment4.html

affords us so promising an opportunity of penetrating the secret of electricity …'

Thomson hypothesised that visible nature of the effects would make it possible to test out ideas on the nature of electricity itself. It was time for electrons to emerge as something far more than units of charge. They might have been around since the Big Bang, but finally they were to enter the spotlight as potential entities in their own right.

THE ELECTRON'S BIRTHPLACE? 17

Cambridge was the place for someone from the Colonies or the Dominions to go on to, and it was to the Cavendish Laboratory that one went to do physics.

Aaron Klug in *Les Prix Nobel* (1982)

Could a fundamental particle have a birthplace? If this is to be a true biography, it's something we have to consider, and though the location I am going to suggest – the Cavendish Laboratory at the University of Cambridge – can certainly be questioned, it's perhaps the best answer we can give.

An initial quibble might be that the Cavendish was not built until the late nineteenth century, while electrons have been around for 13.8 billion years. This is true, though strangely the dating does not disqualify the Cavendish. As it happens, the exact location of the Big Bang was in that laboratory.

This may seem an unlikely coincidence: to be fair, it is also true of wherever you are at the moment while reading this book. This is because in thinking back to the Big Bang we are imagining the distances between points in space getting smaller and smaller as we reverse the expansion of the universe. You can pick any point in the universe and at the point in time that electrons came into being, it was exactly where the Big Bang occurred.

A home from home

More specifically, as I have been writing this book it strikes me how often I have had to take the reader to Cambridge. Of course, there are many other locations where the electron and its works have been studied, but even before the existence of the Cavendish, Cambridge featured regularly. And once the Cavendish came into being, it would play a major role in our understanding of this quixotic particle.

As we will see, the Cavendish Laboratory did not open until June 1874, arguably late in the study of physics. This reflected an early prejudice in academia against the experimental sciences, preferring as much as possible to keep physics within the bounds of mathematics. Of course, experiments had been undertaken since well before Newton's day. But this work had usually been undertaken by individuals in personal laboratories, often little more than a shed or side room in a house. Bringing experiments into the mainstream of professional physics was a Victorian undertaking.

At Cambridge, the driving force behind the establishment of a physics laboratory was the university's chancellor, William Cavendish,* the seventh Duke of Devonshire. When a student at Cambridge himself, Cavendish had been a respectable mathematician, and in the 1870s he funded the post of Cavendish Professor of Experimental Physics and the Cavendish Laboratory. On 9 February 1871, a Grace of the Senate of the University of Cambridge announced that 'the principal duty of the professor is to teach and illustrate the laws of Heat, Electricity and Magnetism; to apply himself to the advancement of the knowledge of such subjects; and to promote their study in the university'.

Perhaps thankfully, the first two physicists approached – William Thomson and Hermann von Helmholtz – turned the job down, leaving it open to the third choice, James Clerk Maxwell, to accept the offer of £500 a year to take the chair and to establish the laboratory. Maxwell

* William Cavendish's grand-uncle Henry Cavendish was a leading eighteenth-century scientist. Notoriously shy and eccentric, Henry had played a hand in the founding of the Royal Institution, experimentally determined the density of the Earth and hence made possible the calculation of Newton's gravitational constant G, and discovered hydrogen.

proved to be an ideal choice in both roles. He had already been a physics professor (though he had never run a university laboratory), but importantly he also had experience of managing a large estate, which would prove beneficial in overseeing the project of building the laboratory.

This was by no means the first such lab, but there were only a handful in the UK at that time, which Maxwell toured not only to get ideas, but also to see how he could better them. Central to his vision for Cambridge was to have a centre that combined the theory and use of mathematics, which he had been so involved with in his career to date, with an experimental foundation, where much experimenting to date had lacked the rigorous connection to mathematics. Cambridge already had that mathematical expertise – now it would combine this with experimental excellence.

So far, so good – but finding a location for a new laboratory near the centre of the medieval city was a nightmare.* Eventually, the site of the old Botanic Gardens was selected, in the (very) narrow street Old School Lane, so close to the walls of Corpus Christi College on the other side of the street that the college unsuccessfully sued the university for putting it in the shade, causing a 'loss of ancient lights'.

Designing a physics laboratory

The architect, William Fawcett, who had primarily worked on churches, wanted a dramatic Gothic design. Maxwell calmed this down, though the main entrance still had the feel of the entryway to a cathedral close, featuring a decorated archway carrying the Cavendish coat of arms and that of the university, topped by a small statue of the Duke himself. The gateway is inscribed with the Latin phrase *Magna opera*

---

* Recognising the limitations of the location of the Cavendish, the main physics site moved out to the west of the city in the early 1970s to form the New Cavendish. After desultory use there was talk for some while of demolishing at least part of the original building, but the core of Maxwell's original building is to be preserved as the 'Old Cavendish Experience', due at the time of writing to be opened in 2026.

Domini exquisita in omnes voluntates ejus (The works of the Lord are great: sought out of all them that have pleasure therein), taken from Psalm 111.

In a letter to his friend William Thomson, Maxwell provides a wonderful insight into the facilities that would be included in the building:

> Lecture room taken for granted.
> Place to stow away apparatus d^{O}*.
>
> Large room with tables &c for beginners at experiments, gas and water laid on &c.
>
> A smaller place or places for advanced experimenters to work at experiments which require to be left for days or weeks standing.
>
> A place on the ground floor with solid foundations for things requiring to be steady.
>
> Access to the roof for atmospheric electricity.
>
> A place with good ventilation to set up Groves or Bunsens batteries without sending fumes into the apparatus.
>
> A good Clock in a quiet place founded on masonry, electric connexion from this to other clocks to be used in the expts and from these connexion to machines for making sparks, marks on paper &c.
>
> A well constructed oven, heated by gas to get up to a uniform high temperature in large things.
>
> A gas engine (if we can get it) to drive apparatus, if not, the University crew in good training in four relays of two, or two of four according to the nature of the expt.†

Maxwell would build a 180-seater lecture theatre (inspired in scale, perhaps, by the lecture theatre at the Royal Institution, though very different in style). There would be a teaching laboratory with ten

---

* Maxwell often used DIY abbreviations in his letters. Here d^{O} represents 'ditto' – i.e. also taken for granted.

† The idea of using the University (rowing) crew to power experiments is an example of Maxwell's distinctive sense of humour.

workbenches, and a number of smaller rooms where postgraduates could run their experiments undisturbed by the hurly-burly of undergraduate experimental sessions. It was in the detail of designing these Spaces that Maxwell really led the field.

For example, the lab workbenches were cushioned from the floor to prevent vibration from influencing experiments. Metal pipes were always visible so that those undertaking electromagnetic experiments could avoid them, and a vacuum system was available throughout the building. Cast iron hollow bricks were built into the walls at intervals to support apparatus, while ceiling joists protruded through the plaster so that supports could be fixed to them. There were even especially wide external window ledges, still visible on the building, so that heliostats – clockwork devices that tracked the Sun and kept a supply of bright light flowing into the building – could be stationed on them.

The building officially opened in June 1874, laying the groundwork for the exploration of the nature of the electron and other components of the atom. This building, now feeling extremely old-fashioned, was state of the art in its day. Beyond the perhaps frivolous suggestion that the Cavendish is the birthplace of the electron as the Big Bang took place there, it was certainly in the Cavendish that we would see the birth of an understanding of the existence of the particle that we are celebrating here, led by a Mancunian.

A SINGULAR CORPUSCLE 18

The electrified-particle theory has for purposes of research a great advantage over the aetherial theory, since it is definite and its consequences can be predicted; with the aetherial theory it is impossible to predict what will happen under any given circumstances, as on this theory we are dealing with hitherto unobserved phenomena of the aether, of whose laws we are ignorant.

J.J. Thomson in *Cathode Rays, Philosophical Magazine* (1897)

Following Crookes' work, J.J. Thomson would identify the electron as an individual 'corpuscle' (the term particle was not yet widely used). The word 'electron' had already been deployed as a quantity of electrical charge, but now it was understood to be an object in its own right, though Thomson was never happy with its use.

There had been hints of the electron's existence before. Looking back in his Nobel Lecture, given in 1938, J.J.'s son George noted, 'Although [Michael] Faraday does not seem to have realized it, his work on electrolysis, by showing the unitary character of the charges on atoms in solution, was the first step. [James] Clerk Maxwell in 1873 used the phrase a "molecule of electricity" and [Hermann] von Helmholtz in 1881 speaking of Faraday's work said "If we accept the hypothesis that elementary substances are composed of atoms, we cannot well avoid concluding that electricity also is divided into elementary portions which behave like atoms of electricity."'

The youthful professor

As we have seen, it was in the Cavendish Laboratory in Cambridge that the electron would finally emerge from its anonymity to become a player in its own right, worthy of a biography. James Clerk Maxwell's laboratory had been run since 1884 by Joseph John Thomson, universally known as J.J. At the time of his appointment, he was just 28 years old.

The idea of such a young person becoming a physics professor who headed up a major lab would now seem bizarre, but it was common during the second half of the nineteenth century when physics was only starting to be recognised as a significant discipline in its own right. Up to this point, as we have seen, what we would now consider physics had largely been subsumed under mathematics: Newton, for example, was Lucasian Professor of Mathematics at Cambridge.

Certainly, at the time, youth was no barrier to taking on a professorial chair. For example, among other physicists, William Thomson (no relation to J.J.), later Lord Kelvin, became Professor of Natural Philosophy at Glasgow aged 22, and James Clerk Maxwell took up a similar professorship at Aberdeen's Marischal College aged 24.

Although Thomson is inevitably linked with Cambridge in the history of science, he was born in the other great British city of physics during this period: Manchester. From the age of thirteen he attended Owens College, an establishment that would later become the Victoria University of Manchester, and which will feature again in chapter 22 when it hosted another of the electron's godparents. At the time, the college was a unique equivalent of what would now be a University Technical College, a STEM-focused secondary school specialising in science and engineering. It was established with an affiliation to the University of London to give it suitable academic credentials.

When Thomson won a place at Cambridge as an undergraduate, the Cavendish was still shiny and new, having only been open for two years. The 1880s saw fast-moving progression for the young man. He became a university lecturer in 1883, was made a Fellow of the Royal Society in 1884, and took over from John Strutt, Lord Rayleigh (who replaced Maxwell after his tragic early death aged 48 in 1879), as Cavendish Professor in 1884. Thomson himself said that his elevation to the job was a surprise. Although there were more experienced men (inevitably)* on the shortlist, some have suggested Thomson's appointment was an intentional break from earlier generations, with some weight behind the choice coming from his powerful college, Trinity.

Apparently, when one of the competitors for the post, Arthur Schuster, based in Manchester, complained to a more senior scientist, he was told, 'Well, no one can say the appointment was a bad one ... it might perhaps have been a better one.' Although Thomson had initially shown more interest in physical chemistry than electricity, the early 1880s found him undertaking his first serious experimental role on the way an electrical coil could act as a condenser (capacitor in modern language – a storage component for electrical charge), but he moved on to more important work in confirming one of Maxwell's most important observations.

Towards a novel particle

As we have seen, in constructing a mathematical model of electromagnetism, Maxwell had noted that the ratio of two units of electrical charge – the electrostatic and the electromagnetic – had the dimensions of speed and had a value similar to that of the (then roughly) measured speed of light, leading to his conclusion that light was an electromagnetic wave. But at the time, neither value had been particularly accurately measured. In 1883, Thomson came up with a value for charge ratio that was extremely close to the then best measurement of light speed.

* It was no surprise only men were on the shortlist for Cavendish Professor. In Thomson's day, women were only starting to gain access to maths and physics at Cambridge. At the time, when women appeared in the Mathematics Tripos listings – the league table of best mathematicians of the year – they weren't allowed the title of 'Wrangler' but instead would be classed as, say, 'between the 9th and 10th Wranglers'. When, in 1890, Philippa Fawcett came top, she was listed as 'above the First Wrangler'. The first women were allowed onto the Natural Sciences physics course during Lord Rayleigh's tenure as Cavendish Professor, just four years before Thomson's appointment.

Thomson took Crookes' work with vacuum tubes much further, thanks to a combination of better vacuum pumps to remove more confounding gas molecules, and making use of a wider range of electromagnetic influences – he introduced extra positive and negative electrodes to study their impact on the cathode rays, as well as making use of magnets to deviate the path of the cathode ray beam (just as those cathode rays would be manipulated in the tubes used in old-fashioned TVs). Thomson's experimental setup was definitive, enabling him to calculate both how fast the components of the rays were travelling and the mass of the particles that appeared to make them up.

It's worth taking a brief step back from Thomson's work to get a better picture of him as a working scientist. We have a useful pointer in a letter from the 1890s that the then youthful physicist Ernest Rutherford wrote back to his fiancée in New Zealand. Rutherford noted, 'I went to the lab and saw Thomson and had a good long talk with him ... As regards appearance he is a medium sized man, dark and quite youthful still, shaves very badly and wears his hair rather long. His face is rather long and thin; has a good head and has a couple of vertical furrows just above his nose ... He asked me up to lunch to Scroope Terrace* where I saw his wife, a tall dark woman, rather sallow in complexion, but very talkative and affable.'

Underlying the development of Thomson's thinking on the Crookes tube was a major shift in his conception of atoms – we have to remember that atoms were mostly thought of as explanatory models rather than real physical particles at this time. It was arguably not until Einstein's work in 1905 that atoms-as-particles were firmly established. Thomson had followed his unrelated namesake William Thomson, Lord Kelvin, in considering atoms to be vortices in the aether – interacting stable rings, which Kelvin thought might form a

* Scroope Terrace was the location of Thomson's home at the time, now backing onto the James Dyson building, part of the Engineering Department of the University of Cambridge. In 1899, the family moved to a significantly larger detached house in West Road.

variety of knots, corresponding to atoms of the different elements.* These knotted vortex rings were not entirely dissimilar to the much later string theory, though here no aether is required.

However, J.J. Thomson quietly dismissed Kelvin's vortex theory to think of atoms more as particles, carrying electrical charges when they were ionised, making it easier to also think of electrons as charged particles in their own right. Having said that, for some time Thomson was inclined to consider Faraday tubes instead. Unlike Crookes tubes, these weren't physical objects, but rather vortex-like structures in the aether that Thomson devised, based on Faraday's lines of force, to provide a link between atoms with opposite electrical charges. However, although it took him a long time to entirely dismiss the aether, his experiments would carry his thinking in a different direction.

As Thomson pointed out in his article titled 'Cathode Rays' for *Philosophical Magazine* published in October 1897, just as we saw when Crookes was experimenting with his tubes, theories for the nature of cathode rays were broadly split between 'some process in the aether' and the wholly material, marking 'the paths of particles of matter charged with negative electricity'.

Though experiments to discover the existence of the aether repeatedly failed to detect any effect, and Maxwell's determination that light was an interaction between electrical and magnetic fields did away with a need for the aether, Thomson was working at a point when the existence of the aether was still clinging on as a possibility. It would take Einstein's special theory of relativity in 1905 and its precursors to do away with any possibility of the existence of such a material.

Consequences of the electrified particle theory

In his 1897 article, Thomson described experiments 'made to test some of the consequences of the electrified particle theory'. He started by showing that cathode rays were one and the same as electrically charged particles emitted from the cathode. This was to clear up, as

---

* The now rich mathematical field of knot theory was effectively started by William Thomson in pursuance of his atoms-as-knots hypothesis.

he put it, the alternative where 'the supporters of the aetherial theory do not deny that electrified particles are shot off from the cathode; they deny, however, that these charged particles have any more to do with the cathode rays than a rifle-ball has with the flash when a rifle is fired'. He did this by placing an electrical detector in an evacuated tube away from the path of cathode rays. This detected very little, which would not be the case if electrons were spraying out in all directions.

Thomson then bent the cathode rays magnetically so rays that were initially hitting the detector were moved away from it – the detector picked up negative electrical charges when the cathode rays hit it, but not when they were deflected to either side. He commented that 'however we twist and deflect the cathode rays by magnetic forces, the negative electrification follows the same path as the rays, and this negative electrification is indissolubly connected with the cathode rays'.

Thomson was also able to dismiss another objection raised by the aethereal theory supporters – they had found that a moderate electrical field would not change the path of cathode rays, suggesting that cathode rays were more like light. However, Thomson surmised that the problem might have been there was too much gas left in the tube in earlier experiments and that as a result the cathode rays were making the gas a conductor (by turning atoms into charged ions), which meant the electrical charge was leaking away.

Because the Cavendish had better vacuum pumps than those used in earlier experiments, Thomson was able to take more of the gas out, providing less material to become conductive – and now was able to deflect the cathode rays with as little as two volts of electricity. He cleverly noted how the quality of the vacuum had an effect by reducing the pressure of the gas to a stage where deflection just began. 'At this stage there is a deflexion of the rays when the plates are first connected with the terminals of the battery, but if this connexion is maintained the patch of phosphorescence gradually creeps back to its undeflected position.'

The phosphorescence he referred to was the green glow produced at the far end of the glass tube when the cathode rays hit it, showing whether or not the rays were being deflected. Initially, the deflection occurred, but with time, what remained of the gas in the tube was becoming ionised and starting to conduct, preventing the weak electrical field from influencing the path of the rays.

After determining that the direction of the rays was not affected by having different gases in the tube, but was under the influence of both electricity and magnetism, Thomson was able to conclude, 'I can see no escape from the conclusion that [cathode rays] are charges of negative electricity carried by particles of matter. The question next arises, What are these particles? are they atoms, or molecules, or matter in a still finer state of subdivision?'

To make the distinction between atomic-scale particles or smaller ones he proposed to measure the ratio of the mass of the particles to the charge they carried. It was easy enough to measure the charge on a stream of particles using an electrometer,* the device he had already used to detect their presence. He then proposed to measure the increase in temperature of a solid body that was being struck by the beam of particles – this would be result of the kinetic energy of the particles being converted into heat.

A final measure would be curvature of the particles in a magnetic field, which should provide a measure for the momentum of a particle divided by its electrical charge. By combining these three measurements it would be possible to calculate both the velocity of the particles and the ratio of their mass to their charge. He also used a second, more complex means to come to the same values to be able to home in on relatively accurate numbers.

The result was a value for the mass to charge ratio of around 10^{-7} (1/10,000,000), which Thomson noted was 'very small' compared with 10^{-4} (1/10,000), which was the value obtained for a hydrogen ion, the lightest known ion (which we now know to be a proton).† One oddity about Thomson's article to the modern reader is that he

---

* Electrometers were the next stage from electroscopes, which merely detected the presence of electrical charges, where a pair of gold leaf flaps move away from each other due to repulsion when charged. Thomson was probably using a Kelvin electrometer, which measures the degree of deflection of a small piece of aluminium reacting to the charge on plates above it.

† It's often the case that a number of scientists come to a similar discovery around the same time. Weeks before Thomson's publication, both Emil Wiechert and Walter Kaufmann had come up with a similar measurement for *m/e*, but it was only Thomson who used his results as evidence that cathode rays were made up of sub-atomic particles – electrons.

doesn't give any units for his measurement. It doesn't matter as far as the comparison goes, but it still seems strange.

Strictly, this doesn't tell us that what we have is a particle with less mass than an atom. As Thomson tells us, 'The smallness of *m/e* may be due to the smallness of *m* or the largeness of *e* or a combination of these two.' However, Thomson asserted that the most likely explanation was that electrically charged corpuscles were components of molecules, giving the same *m/e* whatever gas, for example, was in the tube. Having said that, he missed initially the equivalence of the electron's charge to that appearing on a simple ion, saying, 'There seems to me some evidence that the charges carried by the corpuscles in an atom are large compared with those carried by the ions of an electrolyte,' where in fact they are same.

Thomson would also write, '[from previous experiments] it follows on the hypothesis that cathode rays are charged particles moving with high velocities, that the size of the carriers must be small compared with the dimensions of ordinary atoms or molecules. The assumption of a state of matter more finely subdivided that the atom of an element is a startling one ...' The smallness (in mass) of the electron was primarily down to speculation to fit it into his developing picture of the world at particle level, rather than any evidence from the ratio he had measured – it would take several years of measurements from a number of sources to confirm the electron's tiny nature compared with even the smallest atom.

A revolutionary viewpoint

Because we now look at the electron as a familiar concept, it can be easy to underestimate how revolutionary this discovery was in our understanding of matter. Thomson had shown that the electron had a far smaller mass to charge ratio than the lightest ion – more than a thousand times less than hydrogen. Atoms were supposed to be the smallest, undividable unit of matter, yet here was something likely to be far smaller in mass than the smallest atom – the 'indivisible' atom seemed to have constituent parts. Thomson did try to put this into context by saying that 'a hypothesis that would involve somewhat similar consequences – viz. that the so-called elements are compounds of some primordial element – has been put forward from time to time by various chemists'. But that was mostly thinking of heavy elements being made up of lighter ones, rather than something so radically different as particulate electrons existing within atoms.

If the formation of matter shortly after the Big Bang could be seen as the birth of the electron at the beginning of its biography, this is arguably the point that the particle was recognised – you might say, the particle's coming of age. In describing scientific history, just as much as biographies, it is easy to emphasise something with hindsight that did not have as much significance at the time, in part because it is very rare that a single person, working alone, is the only one to be thinking about and working on a particular issue. American professor of history and philosophy of science Theodore Arabatzis has commented that the 'notion of the "discovery" of the electron has become a problem demanding an explanation, rather than an explanatory resource'.*

That assertion needs some unpacking. Arabatzis is concerned both with what is meant by 'electron' – which as we have seen went through significant changes since the term was first used – and for that matter 'discovery' itself. It is easy to see this as unnecessary academic befuddlement concerning a term that most of us understand perfectly well, but there is a point when considering something like a quantum particle, as the electron would later prove to be. Not only had no direct observation of a particle been made, the very understanding of what a particle *is* would change dramatically in the scientific revolution that occurred in the decades following Thomson's work. But this misses the point. If we are to identify a point in time when the basic concept of an electron as a distinct particle crystalised, it surely is perfectly acceptable to consider this to be the outcome of Thomson's publication.

In passing, I ought also to say that in attributing this work to Thomson I am applying the convention of the time where the principal investigator took most of the glory. But even as far back as the 1890s, much physics was not a solo activity. To give Thomson his

---

* The assertion about the discovery of the electron being a problem demanding an explanation comes in *Representing Electrons*, Theodore Arabatzis (University of Chicago Press, 2005).

due, he finishes his article with the unusual accolade: 'I have much pleasure in thanking Mr Everitt for the assistance he has given me in the preceding investigation.' This references Ebenezer Everett (the spelling seems to be an error in the magazine), who was Thomson's private assistant for 42 years between 1888 to 1930. According to Everett's obituary, written by Thomson in 1933, Everett was a very skilled glass blower (essential for the kind of apparatus used here) and 'he made all the apparatus used in my experiments for the more than forty years in which he acted as my assistant'.

As we saw in chapter 3, the chemist George Stoney had come up with the term 'electron' to describe the size of the minimal charge of an ion in 1891. It quickly became popular after Thomson's discovery to refer to the particle he had made measurements on with this name, though Thomson himself was not particularly fond of the term, as initially it was used for both negative and positive charges, so he stuck with his preferred 'corpuscle' to describe what we now call the electron for nearly twenty years.

From electrical discharges to atomic components

Thomson would go on to write two further papers in 1898 and 1899 which collectively reinforced his argument not only that cathode rays were made up of a stream of corpuscles, but that these particles did indeed feature in the structure of atoms. This was supported by finding the same *m/e* ratio for other electric phenomena, notably the flow of electricity in the photoelectric effect (see page 157) and the electrical discharge given off by hot electric filaments like those in a light bulb.

By December 1899, *in On the Masses of the Ions in Gases at Low Pressures*, Thomson was able to write, 'These results, taken in conjunction with the measurements of the negative ion, suggest that the ionisation of a gas consists of the detachment from the atom of a negative ion; this negative ion being the same for all gases, while the mass of the ion is only a fraction of the mass of an atom of hydrogen … indeed, it seems not improbable that [this negative ion] is the fundamental quantity in terms of which all electrical processes can be expressed.'

Physicists were quick to accept that electric charge was carried by an 'electric atom' – a corpuscle to Thomson – and by 1899 it was already referred to by many as an electron. But we need to underline again that what was far less agreed at this stage was whether or not electrons were interior components of the atoms that made up matter. The very word 'atom' derives from the Greek *atomos*, meaning 'uncuttable'. As the historian of physics Isobel Falconer put it, 'A divisible atom smacked of alchemy.'*

If an atom had component parts, it was argued, then surely the pseudo-scientific concept of transmutation of elements should be possible. The alchemists believed that this was achievable – for example transmuting lead into gold. By Thomson's day, this had become as unlikely as astrology. It would take the development of atomic physics for it to become clear that transmutation was indeed possible, though not produced by the kind of chemical processes attempted by alchemists. To some it was Thomson's tying together of the electrical corpuscle and the atomic constituent that shows why his contribution among work by a whole host of other physicists at the time (including George Johnstone Stoney, Philipp Lenard, Pieter Zeeman, Hendrik Lorentz and George FitzGerald) should give him the title of the electron's discoverer.

Others have suggested† that the most significant aspect of Thomson's contribution to the understanding of electricity was his realisation that electrical currents usually involve the flow of negative charges – a lack of symmetry compared with the general assumption since Benjamin Franklin's day that there were both positive and negative flavours of electricity. It's certainly true that Thomson's work was not the most significant in establishing the electron's role and status in atoms – but from our viewpoint it can still be seen as our first real recognition of the existence of the electron, uncovering part of its role in both electricity and matter.

---

* The likening of a divisible atom to alchemy is from 'Corpuscles to Electrons' by Isobel Falconer, part of *Histories of the Electron* (MIT, 2001).

† The suggestion of the importance of the asymmetry of electrical charge in Thomson's discoveries comes in 'J. J. Thomson and the Electron, 1897–1899' by George Smith, part of *Histories of the Electron* (MIT, 2001).

In reality, although we can celebrate Thomson's contribution, it is probably not realistic to describe the electron as having a discoverer per se. This is something that can only be 'seen' indirectly through its actions. It's not like discovering a continent or an animal species. The label of discoverer doesn't make much sense here. Perhaps it would be better to describe Thomson as a godparent of the electron.

Thomson finished off his work at this stage with some speculation on the implications for a stable internal structure of atoms involving numbers of his charged corpuscles, which would later be refined by him to come up with one of the most misunderstood analogies in all of physics: his plum pudding model of the atom.

19

PLUM PUDDINGS

Professor Thomson suggests [that] … while the negative electricity is concentrated on the extremely small corpuscle, the positive electricity is distributed throughout a considerable volume. An atom would thus consist of minute specks, the negative corpuscles, swimming about in a sphere of positive electrification, like raisins in a parsimonious plum pudding, units of negative electricity being attracted toward the centre, while at the same time repelling each other.

Unattributed notes in Merck's Report on J.J. Thomson's *The Corpuscular Theory of Matter* lectures at the Royal Institution (1906)

Although J.J. Thomson appeared to be moving away from his use of Faraday tubes and a dependence on the aether as he gave more weight to his corpuscles, for a number of years both the aether and the tubes played a part in his thinking about the structure of atoms. It's hard now to appreciate just how central the aether was to the vision of late Victorian scientists.

This meant that as late as 1904, Thomson was able to write that the 'mass of a charged particle arises from the mass of the ether bound by the Faraday tube associated with the charge'.* Quite why he felt it

* Thomson's description of the association of mass with the aether comes from *Electricity and Matter*, J.J. Thomson (Scribner, 1904).

was so difficult to imagine electrons as particles that had mass in their own right is not clear (ironically, it is now thought that electrons get their mass from the Higgs field, though this is a very different concept from the luminiferous aether). It was a little like hanging onto the messy complexities of Ptolemy's epicycles, required to deal with explaining the solar system using a geocentric model, after it had been established that the Sun was at the centre of the system, rather than the Earth. As rational as scientists may be, they can struggle to let go of treasured theories.

Initially, Thomson's move away from Faraday tubes and the aether seems to have primarily been a matter of simplification for general audiences, where he would just refer to 'corpuscles' and 'positive electricity' which acted as mortar to hold the bricks of corpuscles together in atoms. He was still envisioning the role of Faraday tubes and the aether, but simplified them under the hand-waving concept of positive electricity.

The proof is not in the pudding

This simplified version of his theories leads to the 'plum pudding model' described in the opening quote of this chapter. The name often causes confusion because no plums were involved in the development of the model. Instead, the analogy was with a Christmas pudding, a traditional English seasonal dessert, where dried fruit, primarily raisins, are suspended in a doughy mix. Such puddings did not contain plums but were described this way from an obsolete usage referring to all dried fruit, whatever its nature, as plums. In this model we substitute electrons for the dried fruit and a tenuous positive region for the dough.

We still tend to associate the plum pudding model with Thomson, yet it was not one that he was enthusiastic to maintain for any length of time. In 1904, he wrote to fellow English physicist Oliver Lodge, 'I have always had hopes (not yet realized) of being able to do without positive electrification as a separate entity and to replace it by some property of the corpuscles. When one considers that all the positive electricity does on the corpuscular theory is to provide an attractive force to keep the corpuscles together with all the observable properties of the atom determined by the corpuscles, one feels, I think, that the positive electrification will ultimately prove superfluous and it will be possible to get all the effects we now attribute to it from some property of the corpuscle.'*

There was no doubt that keeping the electrons together required some kind of force, as these negatively charged particles would inevitably repel each other, though it seems odd not to mention that the overall atom had to have no electrical charge. To balance out the negative input of the electrons, it is necessary to have a positive charge *somewhere*. However, we always have to bear in mind that to Thomson at this stage the electrons did not exist in isolation but rather as a part of a system with the (eventually abandoned) Faraday tubes, from which he could engineer the ability to cancel out the negative charges.

As we discovered in the previous chapter, Thomson assumed his corpuscles had a mass equivalent to around one thousandth of the mass of the simplest atom, hydrogen. This wasn't too far out when compared with modern measurements, which puts a hydrogen atom at about 1,840 times the mass of an electron. If the 'positive electrification' were insubstantial, meaning that all the mass of the atom came from its electrons, this would require the hydrogen atom to contain around 1,840 electrons. We now know the exact number to be one electron. From our privileged viewpoint of looking back on the electron's life story, we can see that something needed to give.

In more detail (and without the complexity of Faraday tubes), what Thomson envisaged was a 'sphere of positive electrification' within which one or more revolving rings of corpuscles (electrons), evenly spaced, existed. This, he believed, would provide a stable structure which kept the electrons in place despite their tendency to repel each other.

Thomson was still a firm believer in the existence of the aether in 1907, by which point Einstein had already published his paper on the photoelectric effect. As we will discover in the next chapter, this

* Thomson's hope to do away with separate positive electrification is from a letter written to Oliver Lodge on 11 April 1904, quoted in *The Life of Sir J. J. Thomson, sometime Master of Trinity College, Cambridge*, Robert Strutt (Cambridge University Press, 1942).

established that light acted as a stream of particles, the paper among his 1905 *annus mirabilis* papers that would later win him the Nobel Prize. The aether had originally been introduced as a medium for light waves in an otherwise empty space – but if light was a stream of particles, no such medium was required.

Despite the plum pudding description, there is no doubt that Thomson was at least toying with the existence of positively charged particles in atoms. He spent a considerable amount of time studying so-called positive rays in discharge tubes and by 1909 was at least beginning to question whether positive electricity was just as granular as negative: 'Is positive electricity molecular in structure? Is it made up into units, each unit carrying a charge equal in magnitude though opposite in sign to that carried by a corpuscle?'*

As electricity was revealing its secrets, so the nature of the atom was also being explored. Thomson's model of an atom with negative charges spread through a positive goo, like fruit in a Christmas pudding, would be disproved by Ernest Rutherford's striking experiment.

The energetic New Zealander

Rutherford was a very different character to Thomson. Born in New Zealand in 1871, he had an early interest in science (his mother claimed he read Balfour Stewart's *Primer of Physics* aged ten). By 1894, he had three scientific degrees from Christchurch College, Canterbury, part of the University of New Zealand. To further his career, he applied for an 1851 Exhibition Scholarship to start postgraduate study at Cambridge in England.

Like London's Imperial College, the Albert Hall and the Natural History, Science, and Victoria and Albert museums, the 1851 scholarship fund was a recipient of the healthy profits from the 1851 Great Exhibition at Crystal Palace in London. Rutherford was up for the single place allocated to New Zealand against the chemist J.C.

---

* Thomson's speculation on positive electricity 'molecules' is from his presidential address in the *Report of the British Association for the Advancement of Science* (1909).

Maclaurin – and lost out. But love got in the way of Maclaurin's plans – he got engaged, and the award could not be given to a married man, leaving Rutherford to take the prize. According to Rutherford legend, he was digging potatoes on the family farm when the news came through. He is said to have thrown his spade in the air, shouting out, 'That's the last potato I'll ever dig!'

Rutherford's studies took him to Cambridge (where he would eventually return to take over from J.J. Thomson as Cavendish Professor in 1918), then to Canada in 1898, where he undertook the work on isotopes and radioactive substances that would win him the 1908 Nobel Prize in Chemistry. The year before becoming a Nobel laureate he had moved to Manchester in the north of England where, with two of his graduate students, Hans Geiger and Ernest Marsden, he would carry out the experiment that blew the plum pudding model apart.

The experiment made use of alpha particles (which we now know to be helium nuclei), positively charged particles given off by some radioactive materials – in this case, radium. These particles were directed towards a thin piece of gold foil, with the path of the particles followed by placing phosphorescent screens around the foil that would produce a flash when an alpha particle hit them. Geiger or Marsden had to crouch close by in a wooden box to keep things dark and look out for the flashes to pin down the direction the particles had taken.

The assumption was that on encountering the fuzzy positive electrification of Thomson's pudding, the positively charged alpha particles would be slightly deflected off their original course. But to everyone's surprise, some of the particles were deflected back the way they came. Rutherford commented that this was like firing a 15-inch artillery shell at a piece of tissue paper and having it come back and hit you.

Although Rutherford was not primarily interested in atomic structure, the outcome of this experiment was the realisation that the positive charge in atoms was not an ill-defined distribution within which the electrons rotated. Instead, the positive charge seemed to be concentrated in the centre of the atom – what would become known as the nucleus, after the term used in biological cells. Some alpha particles were bouncing back off this small, dense body.

It was easy to jump from this discovery to the picture many still have of the atom, as being like a miniature solar system, but as we will discover in chapter 23, this was far from the case. Although not an immediate certainty, the existence of the dense nucleus also provided a second source of mass inside the atom. No longer did its mass need to come entirely from electrons: most would prove to be located in the nucleus.

Rutherford was not looking to come up with a new atomic structure, but his discovery of the nucleus forced it on the scientific world, along with a new challenge of how an atom with a central nucleus and orbiting electrons could possibly be stable. Rutherford was also responsible for allowing another great physicist of the day to start to uncover how the electron's unlikely behaviour made this possible. But before we take on the nature of electrons in the atom, let's take a moment to see how Einstein used electrons to make clear the true nature of light.

20 EINSTEIN'S NOBEL IDEA

1910, by Wilhelm Ostwald; 1912, by Clemns Schaefer; 1912, by Wilhelm Wien; 1912, by Ernst Pringsheim; 1912, by Wilhelm Ostwald; 1913, by Bernhard Naunyn; 1913, by Wilhelm Ostwald; 1913, by Wilhelm Wien; 1914, by Orest Khvol′son; 1914, by Bernhard Naunyn; 1916, by Felix Ehrenhaft; 1917, by Arthur Haas; 1917, by Emil Warburg; 1917, by Pierre Weiss; 1918, by Stefan Meyer; 1918, by Felix Ehrenhaft; 1918, by Max von Laue; 1918, by Edgar Meyer; 1918, by Emil Warburg; 1918, by Wilhelm Wien; 1919, by Svante Arrhenius; 1919, by Max von Laue; 1919, by Edgar Meyer; 1919, by Max Planck; 1919, by Emil Warburg; 1920, by Niels Bohr; 1920, by Wilhelm Julius; 1920, by Heike Kamerlingh Onnes; 1920, by Hendrik Lorentz; 1920, by Leonard Ornstein; 1920, by Wilhelm von Waldeyer-Hartz; 1920, by Emil Warburg; 1920, by Pieter Zeeman.

Albert Einstein's Nobel Prize nominations before 1921

In 1905, Albert Einstein wrote the paper that would win him the Nobel Prize in Physics. This was the year that he developed the theory that effectively proved that atoms exist, but it wasn't for that. It was also the year that he published his special theory of relativity, and a short extension paper to this that introduced $E = mc^2$ to the world.*

* Strictly, Einstein's 1905 paper on the relationship between energy and matter did not introduce $E = mc^2$ to the world, but rather $m = L/V^2$, as the symbols we use now had yet to be standardised.

It wasn't for that either. His Nobel Prize, awarded surprisingly late in 1921, was for explaining the photoelectric effect (and only then, as you can see above, after very many failed nominations).

Despite having died over 70 years ago, in 1955, Einstein remains the most famous scientist worldwide. I asked history professor Michael Gordin if he could explain this. He replied, 'Honestly, this remains a mystery. Historians have developed a reasonable set of explanations for why Einstein became so famous in 1919, but the persistence of that fame after his death is a complex problem and resists a single account. Surely the range of his commitments in the political and philosophical sphere, alongside the truly exceptional import of his contributions to almost every area of fundamental physics, plays a role. It seems hard to deny that Einstein's charisma – he was photogenic and a master of the quotable line – continues to exert a strong appeal.'

The legend and the man

Einstein's genius remains unquestioned, but his fame outside the scientific community, just like that of Newton, seems to be based as much on a legend as it is on the real man. He was superficially affable, happy to discuss his theories with anyone, but he seems to have held back from real in-depth friendship and rarely managed to work effectively with others. Perhaps this lack of trust in others resulted from the sad failure of his first marriage, a marriage based on a love that he never truly felt again.

Einstein was born on 14 March 1879 in southern Germany, in the city of Ulm. The Einsteins seem to have provided a happy home for Albert and his younger sister Maria – known as Maja – but the family finances were rarely stable. Outside the mostly relaxed atmosphere of the family home, Einstein found late-nineteenth-century German culture oppressive. This came from a combination of a distaste for authority and being told what to do and to think, and the rigidity of the education system of the time, where there was no room for flexibility of thought and method. It didn't help that his pacifist views would become increasingly at odds with the growth of a militaristic society.

After teenage years spent growing further away from both school and his country, at the age of sixteen, Einstein made it a personal goal to cease to be a German citizen. In German-speaking Zurich there was an ideal place to get a real education, the Federal Technology Institute (in German the Eidgenössische Technische Hochschule), known as the ETH, where he got a place on his second attempt. Here too Einstein had trouble with authority figures. The head of the physics department, Heinrich Weber, once said to Einstein, 'You're a very clever boy, but you have one big fault: you will never allow yourself to be told anything.' But his time at the ETH was thrilling academically.

Once he'd graduated, Einstein found a job teaching before taking on an unlikely role thanks to his college friend Marcel Grossman. The head of the Swiss Patent Office, Friedrich Haller, was a friend of Grossman's father. Einstein first contacted Haller at just the right time, when a new job was about to be advertised. Haller interviewed Einstein, and though it was obvious that his knowledge was more theoretical than applied, he took the young man on as Patent Officer (third class).

The job was a godsend: Einstein was amazed how easy the work proved. He had never thought of himself as practical. While still at school, he had written in an essay that he imagined he would become a teacher in the theoretical sciences because he had a 'disposition for abstract and mathematical thought' and a 'lack of imagination and practical ability'. It was here, in his spare time, that his remarkable papers of 1905 would be written. From the electron's viewpoint as much as that of the Nobel committee, his masterpiece was his paper on the photoelectric effect.

The luminous generator

Today we have sophisticated photoelectric cells, powering everything from garden lights to solar farms and space stations, but even back in the early 1900s it was known that shining a bright light on some metals would produce a small amount of electricity. We now know that the energy from photons of light hitting the metal was donating so much energy to electrons in the metal atoms that some were blasted out to flow as an electrical current. Although light is insubstantial, each photon is a packet of energy, which can result in electrons gaining sufficient energy to break free from the attraction of the atom's nucleus.

Although at the time the exact process was unclear, it didn't seem surprising that the energy of light that could also heat things up (think a sunny day) would be able to stimulate the production of electrical energy. But there was something strange about the process. In 1902, the Hungarian physicist Philipp Lenard had found an odd result in his experiments. It didn't matter how bright or dim the light was, electrons knocked out by light of a certain colour had the same amount of energy. Move further down the spectrum towards infrared and you would hit a colour where no electrons were produced at all. You would expect, if light were just a wave, as was generally assumed, that the brighter the light you shone onto the metal, the more energy was available to displace electrons. This would see light acting just as you might expect a bigger wave to displace more sand as it crashed onto the beach. But it wasn't happening.

To make matters worse, theoreticians also faced the ultraviolet catastrophe, a problem with existing theory that had made the older German physicist Max Planck suggest that light seemed to act as if it came in packets of energy, later called quanta, rather than in the continuous stream of a wave. Existing theory had predicted that the radiation from a black body* in thermal equilibrium would get more and more energetic as the wavelength of light decreased, leading to the emission of an infinite amount of energy – something that clearly didn't happen.

Though Planck did not believe that his quanta truly existed, they patched up the theory, predicting instead the exact behaviour seen in black body radiation. Einstein took Planck's device to prevent the impossible happening and imagined what would occur if quanta were real. If light acted not as a wave but as a stream of packets – in effect, particles – and each electron that was boosted out of a piece of metal in the photoelectric effect did so as a result of the impact of one of these light particles, then you would expect there would be a cut-off level, below which light quanta with lower energy could not contribute

* In physics, a black body is a theoretical object that will absorb all electromagnetic radiation that hits it, and when heated emits radiation entirely dependent on its temperature.

to an electrical current, however many of them there were – exactly what was observed.

Somehow, Einstein thought, despite all the existing evidence that light could act like a wave, it genuinely was made of these packets of energy. It was left to American chemist Gilbert Lewis to give these packets the name we now use – photons – but Einstein's theory explained away the problems of understanding the photoelectric effect. In this one step, thanks to the foundations Planck had laid, Einstein threw away one of the most basic accepted facts in science, that light was a perfectly normal wave. This was a fundamental step in the foundation of quantum theory. It showed that although light could behave as if it were a continuous wave, some of its nature required it to be thought of as a stream of particles – quanta. And the photoelectric effect connected these light quanta, photons, with the particles central to this book, electrons.

The photon–electron dance

It would not be until the 1950s and 60s, with the development of quantum electrodynamics (which we will uncover in chapter 28), that the full nature of what was happening when light and electrons interacted would be understood. But the interplay of these quantum particles would be not only central in technology to the development of solar panels, but also in digital cameras and more.

These applications, though, are only minor examples of the ways that light adds energy to electrons – or that electrons can lose energy in the form of light. These processes are happening all the time around us. For example, when we see an object there are multiple interactions between photons and electrons taking place.

Let's say you see a red apple. Light will have come from a source, be it the Sun or artificial lighting, and hit the surface of the apple. The photons of light would give boosts of energy to electrons orbiting the atoms in the apple's skin. Soon after, those electrons would lose some energy, emitting new photons, which could have a different energy to the incoming particles, depending on the nature of the atom. In the case of the apple, those photons would primarily have the energy associated with the colour red. The photons would then hit your eye where, in the retina, their energy is converted once more to give electrons a boost, carrying signals to your brain.

When Einstein came up with his theory, relatively little was still known about electrons, or how electrons were structured in an atom. As we have seen, J.J. Thomson had laid the groundwork for the possibility that electrons existed – and Einstein's theory was dependent on this concept. But there were still holes in the evidence, and confusion over how atoms containing electrons could be stable. In the next three chapters, we will see how a drop of oil provided more information on the electron, how electrons could become irresistible, and how a young Danish physicist would be inspired to imagine the unimaginable in the behaviour of electrons around atoms.

MILLIKAN'S OILY DROP 21

We have learned a lot from experience about how to handle some of the ways we fool ourselves. One example: Millikan measured the charge on an electron by an experiment with falling oil drops, and got an answer which we now know not to be quite right. It's a little bit off because he had the incorrect value for the viscosity of air. It's interesting to look at the history of measurements of the charge of an electron, after Millikan. If you plot them as a function of time, you find that one is a little bit bigger than Millikan's, and the next one's a little bit bigger than that, and the next one's a little bit bigger than that, until finally they settle down to a number which is higher.

Why didn't they discover the new number was higher right away? It's a thing that scientists are ashamed of – this history – because it's apparent that people did things like this: When they got a number that was too high above Millikan's, they thought something must be wrong – and they would look for and find a reason why something might be wrong. When they got a number close to Millikan's value they didn't look so hard. And so they eliminated the numbers that were too far off, and did other things like that. We've learned those tricks nowadays, and now we don't have that kind of disease.

Richard Feynman in *Surely You're Joking, Mr. Feynman!* (1985)

While J.J. Thomson had established the likely corpuscular nature of electricity, he had not pinned down the electrical charge carried by an electron, a value that would prove to be a fundamental constant of nature.* Establishing this value, giving further evidence for the existence of the electron as an entity, would be the work of American physicist Robert Millikan. You might imagine that no one would doubt the final validity of this discovery, but Millikan, in his Nobel Lecture, quoted an (unnamed) 'prominent literary writer' of the 1920s who 'spoke of the electron as "only the latest scientific hypothesis which will in its turn give way to the abra-ca-da-bra of tomorrow".' We may have refined our view of the electron since, but it stubbornly remains as a seemingly permanent abra-ca-da-bra.

Inside the abra-ca-da-bra

Born in 1868 in Morrison, Illinois, Millikan initially worked on light polarisation while undertaking his PhD at Columbia University. After a year out in Germany, he got a post-doctoral placement at the University of Chicago, where he would become Professor of Physics in 1910, just one year after he made his name with his experiment to measure the charge on the electron. This was a crucial point in filling out the biographical picture of the electron, as Thomson had only identified the ratio of charge to mass – once the value for charge was identified it immediately made the mass available too, clarifying any potential confusion over the scale of the electron compared with an atom.

Millikan started by following up a series of experiments on electrically charged water drops undertaken by the Scottish inventor of the cloud chamber, Charles Wilson (we will find out more about cloud chambers in chapter 27). Millikan refined the experiment down to a single water drop and then moved to oil because of concerns about evaporation. The principle behind Millikan's experiment, undertaken with his student Harvey Fletcher (who later pioneered stereo sound),

* Arguably, it is more accurate to describe the electron's charge as fundamental-ish than strictly fundamental. Although the electron is a fundamental particle, the equivalent charge in the positive proton is made up of two smaller values on its component quarks. If the electron's charge is -e, the up quark's charge is 2/3 +e and the down quark's charge is 1/3 -e – so there is a particle with a smaller negative charge.

was to balance out the impact on an electrically charged oil drop of electrical attraction and gravity.

As is not uncommon in circumstances where a scientist works closely with a post-graduate assistant, there was some controversy over exactly how much input Fletcher made to the work. Such disputes tend to be particularly fierce when the lead scientist wins a Nobel Prize, as Millikan would for this experiment in 1923, but the assistant, who may have done the bulk of the work, does not.

The most infamous example of this was the case of Jocelyn Bell (now Bell Burnell), who discovered pulsars (fast-rotating neutron stars that produce fast-pulsing radio signals) as a postgraduate researcher in 1967. The 1974 Nobel Prize in Physics was awarded to her supervisor, Antony Hewish, but not to Bell. English physicist Fred Hoyle was convinced this was the result of misogyny on the part of the Nobel Committee, though Bell accepted the outcome, commenting, 'I believe it would demean Nobel Prizes if they were awarded to research students, except in very exceptional cases, and I do not believe this is one of them.'

Given the amount of work Bell put into the discovery, not only in building the equipment but in taking and assessing the readings, she was arguably being over-generous. It has been suggested that Fletcher too should have received accolades for the oil drop experiment, especially as documentation suggesting that Millikan had threatened to not award Fletcher's PhD if he insisted on co-authorship on the oil drop paper has emerged since Fletcher's death.

The flight of the oil drop

The experiment, contained in a cylindrical metal container looking something like an industrial pressure cooker, consisted of two horizontal metal plates 16 millimetres (0.63 inches) apart, which had a variable electrical voltage applied across them. This applied a force to any electrically charged droplets located between the plates. A mist of these droplets was created inside the container, by spraying them through a fine nozzle. Just the action of spraying would naturally give a charge to some of the droplets in triboelectric fashion (see chapter 2), though Millikan suggested the charge on the droplets should be varied by a variety of means.

In his Nobel Lecture he described 'changing its charge in a considerable number of different ways; for example, by ionizing the air just beneath it by alpha, beta, or gamma rays from radium and letting the field throw these ions into the drop; by illuminating the surface of the drop itself with ultraviolet light; by shooting X-rays both directly at it and beneath it, etc.'

The space between the plates was monitored with a microscope. Initially, the electricity was not switched on, simply measuring the rate at which drops that passed through a small hole in the top plate fell. Given the density of the oil and the rate at which the drops fell on reaching terminal velocity, it was possible to work out the gravitational force drops were experiencing. Then the electricity was turned on.

At just the right level of potential difference between the plates, a selected drop could be forced to stop in position between plates, the force of gravity exactly balanced out by the force provided by the electrical field. Given this force and the field strength, it was possible to work on the charge on the droplet. Of course, it was not possible to ensure that each drop had only a single electronic charge, but by repeating this many times it became obvious that the drops all had multiples of the same small value, which showed the size of the electron's charge. The result is less than 1 per cent away from the modern value.*

Errors and cherry picking

This sounds impressive, but it does raise some concerns about Millikan and Fletcher's experimental procedures, as the 1 per cent difference is six times bigger than the standard error provided in Millikan's paper. The standard error is a statistical measure that helps indicate the confidence

* The current value for the charge on the electron is $1.602176634 \times 10^{-19}$ coulombs – this is now a value that will never change, as the unit has been defined in terms of the relevant multiple of the charge on the electron. Something similar was done with the speed of light, which is (and always will be) exactly 299,792,458 metres per second in a vacuum, as the metre is defined from this value. It seems a shame that physicists could not have opted for rounder numbers.

we can have in a result – it is suspicious if a value is so much further out than the standard error suggests is likely. As Millikan appeared to have omitted more than half the measurements taken in his paper, it is possible that he was able to make his apparent error smaller. Note that this is a different issue to the one described by Richard Feynman in the quote that opens this chapter – here the problem comes with measurements made by others after Millikan's experiment.

Whether or not there was fudging in the write-up of Millikan's experiment, it was an elegant way to pinpoint a key part of the nature of the electron – if you like, one of the important aspects of its personality – its electrical charge. Without that charge, the electron would not feature in electricity, nor would it be able to function as it does in the atom. Once the oil drop experiment was undertaken there could be no further doubt about the electron as an 'atom' of electricity. As Millikan put it, 'the electron itself, which man has measured ... is neither an uncertainty nor an hypothesis. It is a new experimental fact that this generation in which we live has for the first time seen, but which anyone who wills may henceforth see.'

In an interesting aside, Millikan, possibly due to his close association with the electron, was not happy with the name 'proton', given to the nuclear particle with the equivalent positive charge. He moaned in his Nobel Lecture, 'It is highly desirable that this historically correct, etymologically most suitable, and authoritatively recognized nomenclature [the electron] (see among many others Rutherford's Presidential Addresses at the British Association 1923, Nemst's *Theoretical Chemistry*, last edition, etc., etc.) be retained. When used without a prefix or qualifying adjective, the word electron may signify, if we wish, as it does in common usage, both the generic thing, the unit charge, and also the negative member of the species, precisely as the word "man" in English denotes both the genus homo and the male of mankind. There is no gain in convenience in replacing positive electron, by "proton", but on the other hand a distinct loss logically, etymologically, and historically.'

Very soon after Millikan's experiment (and well before his Nobel Prize), a mind-boggling behaviour of electrons in ultra-low temperatures would be discovered. It would take the later development of quantum theory to understand what was happening, but a Dutch scientist was about to discover that some electrons seemed unstoppable.

UNSTOPPABLE ELECTRONS 22

As you know, very many metals become superconducting below a certain temperature ... the temperature is different for different metals. When you reduce the temperature sufficiently the metals conduct electricity without any resistance. This phenomenon has been observed for a very large number of metals but not for all, and the theory of this phenomenon has caused a great deal of difficulty. It took a very long time to understand what was going on inside of superconductors, and I will only describe enough of it for our present purposes. It turns out that due to the interactions of the electrons with the vibrations of the atoms in the lattice, there is a small net effective attraction between the electrons. The result is that the electrons form together, if I may speak very qualitatively and crudely, bound pairs.

Richard Feynman in *The Feynman Lectures on Physics* (1965)

As often happens when digging into the past of a multifaceted contributor to history such as the electron, we find that it keeps cropping up in roles that weren't consciously attributed to it at the time. In 1911, just two years after Millikan's oil drop experiment, an old-school Dutch physicist would observe electrons behaving most mysteriously, though (as Richard Feynman suggested in the quote above) it would take a long time for the electron's role to be fully understood.

Dr Cool

Heike Kamerlingh Onnes, born in Groningen in the Netherlands in 1853, was professor of experimental physics at Leiden University for

35 years, beginning in 1882. An expert in low temperature experiments, he was particularly interested in obtaining temperatures close to absolute zero – the lowest conceivable temperature. He was able to achieve particularly low temperatures by deploying the Joule–Thomson effect, where a gas is forced through a narrow orifice.

Here the gas is allowed to expand in an insulated environment, so no heat is exchanged with the exterior, a process that cools most gases. This is because there is a transfer of energy from kinetic to potential energy by the process, and, as we will see, temperature is a measure of the kinetic energy of the gas molecules. An essential realisation for managing to cool helium, which would be central to Kamerlingh Onnes' work, was that this process only worked on helium that had already been cooled to temperatures well below room temperature using another gas.

The idea that there should be a minimum temperature had been around since at least the seventeenth century, when the proto-chemist Robert Boyle made considerable efforts to show that there was no such thing as the *primum frigidum*.* At the time, heat was thought of as a fluid known as caloric that flowed from body to body. The concept of *the primum frigidum* was effectively a source of ultimate coldness in the caloric model. If all caloric had left a body, it could not get any colder, producing a lower limit for temperature.

Although caloric proved not to exist, what became known as absolute zero remained useful. As the science of thermodynamics was clarified, it emerged that temperature was a measure of the kinetic energy of atoms and molecules, effectively how much they jiggled about. As temperature dropped, this motion got more and more sluggish until eventually it should reach a state where movement effectively stopped. In practice, due to quantum effects, absolute zero can never be reached, but temperatures very close to it can be achieved.

With the understanding of atomic structure that started with Niels Bohr's work (explored in the next chapter), the energy that was being

---

* Boyle's attempts to deny the existence of an ultimate cold were in the magnificently titled *New experiments and observations touching cold, or, An experimental history of cold begun to which are added an examen of antiperistasis and an examen of Mr. Hobs's doctrine about cold* (1665).

reduced as temperatures fell also included the energy levels of electrons around atoms. But when everything had reached the lowest possible energy, there was nothing lower to achieve. This was a true zero value for temperature, unlikely the arbitrary zeroes found on everyday temperature scales such as Celsius and Fahrenheit.

William Thomson (Lord Kelvin), who was a major contributor to early thermodynamic theory, proposed a new temperature scale using the same sized units as Celsius (or Centigrade, as it was often then known), but starting at absolute zero. In a more familiar scale this is around -273.15°C or -459.67°F. The scale based on absolute zero is now known as the Kelvin scale, and, unlike the others, refers to its units as kelvins, not degrees.

Heading for absolute zero

Attempts had been made to get well below the freezing point of water since the 1750s, but Kamerlingh Onnes would break every record in 1908 when he liquified helium, the ultimate coldness challenge, getting the element down to around 1.5 K (-271.65°C or -456.97°F). This was a remarkable achievement for which he won the Nobel Prize in Physics for 1913. But it was not the most dramatic discovery that he would make. Kamerlingh Onnes' interest was not just in achieving ultra-low temperatures, but in their impact on the physical properties of matter.

With the gradual understanding of the nature of the electron at the time, it was assumed that electrical conduction in a metal was like water flowing through a pipe, with the electrons taking the part of water molecules. As we have seen, the use of a term like 'current' for electricity suggests this, as does the way that the early electronic components we will meet in chapter 24, controlling the flow of electricity, were referred to as valves. Given a reduction in temperature was equivalent to a reduction in kinetic energy, it was widely assumed that the ability of a material to conduct electricity would become more and more limited as temperature fell and the flow of electrons got increasingly sluggish. This would be seen as an increase in electrical resistance, until no electricity flowed at all at absolute zero.

Kamerlingh Onnes had a different theory, suspecting that resistance would drop with temperature, not conductivity, though he never expected it to reach zero. He set out to explore this possibility using the metal mercury (a solid at these low temperatures). This might seem an unlikely choice when metals solid at room temperature such as copper or iron were the usual electrical conductors. But mercury's low boiling point made it easy to distil to remove any impurities that might have interfered with conductivity, and being liquid at room temperature, it was easy to create thin, consistent wires by first putting it into glass capillary tubes.

Kwik nagenoeg nul

Rather than see a smooth change in the electrical resistance of the mercury, Kamerlingh Onnes found that at 4.2 K all resistance disappeared. It's rare that we know the exact time a scientific discovery was made, usually relying on the publication date of a paper on the subject. But from Kamerlingh Onnes' horribly scrawled laboratory notebooks, we can pin down this result to 4pm on 8 April 1911.* Here he wrote, '*Kwik nagenoeg nul*', which translates as 'Quick pretty much nothing', stating that quicksilver (mercury) had no apparent resistance.

Checking a lack of resistance was not trivial with the equipment available at the time. One experiment Kamerlingh Onnes undertook was to set an electrical current flowing through a ring of superconducting metal, which acted as a simple electromagnetic. Its magnetic field could be measured: a conventional conductor would see the current dropping as resistance produced a loss of energy to heat, resulting in reduction of the field strength. But over the several hours he was able to keep his helium liquid, there was no detectable change. By the

* Getting the date of Kamerlingh Onnes' discovery from his notes is not as simple as it seems, as it appears in a notebook dated 1909–10, and later in the same book an experiment is dated 19 May 1910. But he did not get the apparatus required for that experiment until April 1911, so it appears the year was incorrect.

1950s, a similar experiment could be carried out for eighteen months without any reduction in field strength.

As Feynman suggested, although this superconductivity clearly involved the action of electrons, the early workers in the field had no idea exactly what was happening to those electrons to cause this effect. Conventional resistance happens when the electrons carrying electrical current interact with the atomic lattice of the conductor – despite being a solid, this structure is constantly vibrating, interacting with the electrons to slow them down and create heat. The explanation for simple superconductivity (there proved to be more than one kind) would not come until 1957, when American physicists John Bardeen, Leon Cooper and John Schrieffer would devise the theory later known as BCS that won them the Nobel Prize in Physics for 1972.

Cooper had already suggested that pairs of electrons (known as Cooper pairs) could act as if they were a single particle at extremely low temperatures, forced together by the vibrations within a crystal lattice. Quantum physics made it clear that electrons could not be considered to have an exact location: this meant that pairs of the particles could overlap sufficiently to become a single entity, referred to as a condensate. In this state, the electron pairs could float through the conductor's lattice as if it were not there, producing zero resistance.

Since this early work, superconducting materials have been developed that do not require cooling with liquid helium, which is both expensive and difficult to keep cold enough (hence Kamerlingh Onnes' inability to keep his superconducting ring in action). These so-called high-temperature superconductors only require the use of liquid nitrogen, which boils at a toasty 77 K. This is still chilly by normal measures, but liquid nitrogen is far easier and cheaper to produce, even used by high-end caterers to create instant iced food. The holy grail of superconductivity is to get a superconducting material that functions at room temperature, enabling superconductors to be used, for instance, in lossless transmission of electricity.

Despite a number of announcements that have been retracted, room temperature superconductors have not yet been achieved, but conventional superconductors now have a range of practical uses, most often where a very strong magnetic field is required. We see superconductors in use in MRI scanners in hospitals, in particle accelerators like the Large Hadron Collider at CERN and in maglev trains that float above the tracks without physical contact by magnetic repulsion, enabling extremely high speeds to be achieved. All thanks to electrons getting friendly with each other.

Shortly before Kamerlingh Onnes' discovery, Robert Millikan gave us insights into the electron as a stand-alone particle, a component of electricity, which is certainly one of its essential roles. But as we are discovering, the electron is anything but a one-trick pony, and in 1913, a few years after Millikan's work, a young Danish physicist would finally begin to crack the problem of the structure of electrons in atoms that had been unsuccessfully dealt with in the plum pudding model. Along the way, he introduced one of the major oddities of quantum physics.

BOHR'S AMAZING PERFORMING ELECTRONS 23

The first thing Bohr said to me was that it would only then be profitable to work with him if I understood that he was a dilettante. The only way I knew to react to this unexpected statement was with a polite smile of disbelief. But evidently Bohr was serious. He explained how he had to approach every new question from a starting point of total ignorance. It is perhaps better to say that Bohr's strength lay in his formidable intuition and insight rather than erudition.

Abraham Pais in *Niels Bohr: His Life and Work as Seen by His Friends and Colleagues* (1967)

In 1913, the young Danish physicist Niels Bohr published a three-part paper on his atomic model that would lay the foundation of our understanding of the atom and how electrons exist within it.* The papers were published when Bohr already held a lectureship in physics at Copenhagen University, but its genesis was in the north of England, inspired by the new contacts that Bohr would make there.

The chances are that Manchester meant no more to Bohr when he first planned to come to England than Denmark's third city, Odense, would to most people in the UK. His initial focus was Cambridge

* This chapter is based on the article 'Made in Manchester' by the author, published in *Physics World* magazine in 2013: material reproduced with permission.

University and its already famous Cavendish Laboratory, where he would come under the wing of the electron king, J.J. Thomson. Bohr had received a grant from the Carlsberg Foundation* to spend a year studying abroad, and few locations offered more possibilities in experimental physics than the Cavendish.

A great Dane abroad

Bohr, born in Copenhagen in 1885, had very little English as he set out in September 1911 for Cambridge to meet Thomson – something he tried to remedy by working through the Charles Dickens novel *The Pickwick Papers* with a dictionary alongside. But something Bohr did have was clear academic promise. Before he had even gained his doctorate at Copenhagen University on the behaviour of free electrons in metals, Bohr had already won the Danish Academy of Sciences' gold medal and published two papers in the *Philosophical Transactions of the Royal Society*. The track record of Thomson and the Cavendish Laboratory would inevitably have made Cambridge hugely attractive to Bohr, who had in his thesis referred to Thomson as a major figure in the development of atomic theory.

By the time Bohr arrived in Cambridge, Thomson's plum pudding model was under threat as a result of Ernest Rutherford's work in Manchester, which we encountered in chapter 19. But these results – important though they were – did not prevent Bohr from feeling that the Cavendish, with Thomson at the helm, was the place to be. Unfortunately, any collaboration might have gone better had Bohr's first meeting with Thomson not consisted primarily of Bohr bluntly telling the older scientist that a calculation in one of his books was wrong.

After this uncomfortable start, as he described in a letter to the Swedish physicist Carl Wilhelm Oseen, Bohr found Thomson difficult to hold a conversation with – and Thomson was disinclined even to

* Carlsberg later provided Bohr with its prestigious house in Copenhagen, the most impressive feature of which is that it comes with unlimited Carlsberg lager on tap.

read Bohr's doctoral thesis. Rather than sharing ideas with the great man, as he might have envisaged doing, Bohr found himself primarily occupied attending lectures on electromagnetism and undertaking experimental work on cathode rays. In an interview given late in his life, quoted by his friend and biographer Abraham Pais, Bohr admitted, 'The whole thing was very interesting in Cambridge but it was absolutely useless.'

It is useful to compare Bohr's experience with Thomson with that of the man who would later be his mentor, Ernest Rutherford. We saw in chapter 18 how Rutherford described Thomson 26 years earlier. Rutherford also noted when writing to his fiancée, 'He is very pleasant in conversation and not fossilised at all ... I like Mr & Mrs both very much. She tries to make me feel at home as much as possible and he will talk about all sorts of subjects and not shop at all.' Admittedly, Rutherford was a more easy-going and affable character than Bohr, but it seems that by 1911, when Bohr turned up, Thomson had become more than a little fossilised.

In November, just two months after arriving at Cambridge, Bohr visited a friend of his late father's who lived in Manchester. Ernest Rutherford was also invited to the same soirée, and Bohr mentioned to him that he would be interested in coming up to Manchester to discover more about radioactivity, the main focus of work in Rutherford's lab. This move seems to have been cemented a month later when Bohr encountered Rutherford at a ten-course dinner at Cambridge for physics department members past and present, Rutherford having worked for his PhD there between 1895 and 1898 following his move from New Zealand. By March 1912, Bohr was on his way north.

Heading North

When Bohr arrived in Manchester, the young post-doc and his supervisor proved an interesting pairing. Bohr was poles apart from Rutherford's bluff and open persona. Quiet and reserved, the young Dane had to carefully organise his thoughts before speaking, while Rutherford would leap into a conversation with his comments at a moment's notice. When he was thinking, Bohr would withdraw from reality in a way that seemed to leave him as little more than a shell. A colleague, James Franck, once commented, 'His face became empty, his limbs were hanging down, and you would not know this man could even see. You would think that he must be an idiot. There was absolutely no degree of life. Then suddenly one would see that a glow went up in him and a spark came, and then he said: "Now I know."'

This same intensity of concentration would later come through in Bohr's most entertaining encounter with his long-time academic sparring partner, Albert Einstein. In 1948, Bohr was staying at the Institute for Advanced Study (IAS) at Princeton, using Einstein's office, which was vacant because Einstein found it too spacious and preferred to work in the smaller assistant's office. Bohr had asked fellow physicist Abraham Pais to make some notes for him.

This request reflected a practice that Bohr followed through most of his working life – he much preferred to assemble ideas in his head and then dictate, rather than write things down himself. It has been suggested that, apart from his tendency to construct elaborate text in his mind before pouring it out, this was also because his handwriting was so appalling. It is an indication of how bad it was that when Bohr was operating under the assumed name of Nicholas Baker during the Second World War, known as 'Uncle Nick', there was wide dispute over whether a letter from him to Los Alamos used that signature or Niels Bohr.

When thinking, Bohr tended to pace around and repeat a word over and over until he had assembled a complete sentence in his mind. In the instance at the IAS, Pais recounts, the word in question was 'Einstein', and Bohr came up with it again and again as he walked around the office, thinking about one of Einstein's theories. After a while he was standing looking out of the window, continuing to occasionally say, 'Einstein.'

At this point, Einstein quietly entered the room. He had been forbidden to buy tobacco by his doctor, so had decided he could get around this injunction by purloining some from Bohr's pouch. Einstein tiptoed across to the desk to help himself as Bohr continued to mutter 'Einstein' while he peered out of the window, his back to the desk. As Einstein was reaching for the tobacco, Bohr came out with a final, firm, 'Einstein!' and turned to find the subject of his monologue standing mysteriously before him. 'There they were,' Pais commented, 'face to face, as if Bohr had summoned him forth. It is an understatement to say that for a moment Bohr was speechless.'

A different approach to science

In Manchester, though, Bohr was hugely and positively influenced by Rutherford and the way he ran his department. Bohr's time in England had switched his focus from primarily considering classical physics to the new quantum theory – an interest that first developed while at Cambridge but that flowered in Manchester – and Bohr based his own management style, when later running his own teams, on Rutherford's successful blend of personal research and the development and support of a strong coterie of younger physicists.

The Victoria University of Manchester – as Manchester University was known as in Bohr's day – had been established in 1880 as Owens College, the UK's first civic university. The building where Bohr worked, now called the Rutherford Building, was a shiny new facility when he arrived, having opened in 1900 with a state-of-the-art ventilation system that drew outside air over oil baths to remove the soot particles that tainted the air of the smoky industrial city. It was in this building, and in his rented rooms, that Bohr would undertake his crucial work.

After attending a course on radioactivity, Bohr was once again set to work on an experiment – never truly his forte* – in this case studying the absorption of alpha particles (helium nuclei) by aluminium. For a few weeks he spent all day in the lab, most appreciating the time when Rutherford – always more approachable than Thomson had ever been – would come around and the pair could discuss Bohr's observations. But Bohr was enthusiastic to focus more on theory, and after a few weeks had put aluminium aside for the more amenable laboratory of pencil and paper.

---

* Although Bohr was not a great experimenter, there were far worse among theoreticians. It is rumoured that American physicist Robert Oppenheimer once left a poisoned apple for his supervisor, the English physicist Patrick Blackett, so frustrated was he by his experimental failures, though there is no documentary evidence, and there was certainly no involvement of Niels Bohr as shown in the *Oppenheimer* movie. But king of the bad experiment was Austrian physicist Wolfgang Pauli. In the humorously named Pauli Effect, it was said that whenever he was close to a piece of equipment it would fail.

Bohr's initial ideas were on the newly developing (though yet to be named) concept of isotopes, which had been a major feature of Rutherford's early work. Bohr believed that the defining factor for an element was not its atomic mass, but its atomic number – the size of the positive charge on the nucleus – and he worked out how emitting alpha particles or beta particles (electrons) from the nucleus would change the element. At the time, Rutherford himself played down the significance of a nuclear atom, perhaps because his focus was on scattering effects rather than atomic structure, and seems to have discouraged Bohr from taking his thinking any further. Instead, Bohr built on the work of Charles Darwin's grandson Charles Galton Darwin, a significant physicist in his own right, who around that time was attempting to explain the way that alpha particles lost energy when passing through matter.

Darwin had correctly suggested that this energy loss was due to alpha particles interacting with atomic electrons, but had problems bringing his theory into line with experimental data because he assumed those electrons were floating freely inside the substance. Bohr looked instead at the possibility that they were bound via some kind of elastic connection to the nucleus, meaning they should vibrate with specific frequencies. Max Planck's 1900 theory of radiation from an atom had been based on the idea of a minimum amount of energy that could be radiated by an atomic vibration, and in June, Bohr made the connection that the radiation of energy from these electrons could be limited by Planck's constant,* thereby bringing the early quantum theory into the structure of the atom for the first time.

Don't talk about it to anybody

This work seems to have inspired Bohr to apply his undoubted high-power thinking skills to a bigger prize. As he wrote from

* Planck's constant is a fundamental constant of nature that is linked to the energy of a quantum particle. Specifically, the energy of a photon is its frequency (when considered a wave) multiplied by this constant. The constant came out of Planck's initial work on black body radiation that marked the beginning of quantum physics.

Manchester to his brother Harald: 'Perhaps I have found out a little about the structure of atoms. Don't talk about it to anybody ... It has grown out of a little information I got from the absorption of alpha rays.' Harald, two years younger than Bohr, was also a physicist, though perhaps better-known as a footballer, winning a silver medal in the 1908 Olympic Games. (Niels Bohr was also an enthusiastic player of the game.)

Bohr wrote up his first thoughts in Manchester in a draft that he prepared for Rutherford called 'On the constitution of atoms and molecules', the same title that he would later give to his definitive three-part paper. It was here for the first time that anyone had dared take on the implications of Rutherford's nucleus for the structure of atoms.

As we have seen, Rutherford's experiment located the positive charge in the atom in a tiny but relatively massive nucleus. The negative charge was known to reside in electrons. But what would provide a stable structure for those electrons and the nucleus to coexist in proximity? The negatively charged electrons would seem naturally inclined to plunge into the positively charged nucleus. It seemed initially as if there was no way forward. It was possible to imagine that there would be some stable configuration where fixed electrons repelling each other could sit fixed in place around a positive nucleus, but it had been shown some while before that mechanically there was no such stable configuration. And if the electrons moved in orbits, like the planets in a solar system, that too introduced its own apparently insoluble problems.

It had been known since Maxwell's time that when an electrical charge is accelerated, it gives off electromagnetic radiation. This is how, for instance, radio waves are produced, as we discovered in chapter 14. Moving in any kind of normal orbit implies acceleration. This is because acceleration is a change in velocity, which combines speed and direction of travel. A body in orbit is always changing direction, and hence is changing velocity – it is constantly accelerating. Because of this, orbiting electrons would radiate away their energy and crash into the nucleus, rather as a satellite losing energy in a decaying orbit would spiral into the Earth.

Rather than give up at this point, Bohr took the mental leap of developing a hypothesis that he would admit had no 'mechanical foundation' – in effect he assumed that electrons in an atom behaved entirely differently from the expected behaviour of objects in the macro world. When mechanics had proved unable to explain the workings of the atom, it was time to abandon the conventional approach.

Specifically, Bohr called on the thinking that had been introduced with the first, tentative steps into quantum theory. His hypothesis, he commented, was the only one that seemed to offer an explanation for the experimental results confirming the radiation mechanism proposed by Planck and Einstein, the idea that atoms could only radiate energy in fixed packets of energy called quanta.

A new model atom

Bohr envisaged an atom in which there was a relationship between the energy of an electron and the frequency of its orbit. If you imagine an orbiting electron having an associated frequency, just like the photons that started the quantum theory, then it was possible to imagine a similar relationship for the electron, where the orbit had to be made up of a complete multiple of the particle's wavelength, making it quantised: it couldn't take any value, just those that fitted in this way.

Overall, Bohr's insight while in Manchester was simple but stark. There was no hope, he reasoned, of understanding atomic structure using classical physics. It could only be stabilised by taking a quantum approach that throws away the behaviour we expect from orbiting bodies.

The final piece in the picture had to wait until after Bohr left Manchester in July 1912, when his grant ran out. This involved returning to Copenhagen, first to get married and then to work at the university, where he went on to take up a lectureship in physics the following year. Although the atomic model Bohr devised in Manchester had a quantised orbit, it was still unstable because it was capable of radiating its way into the nucleus. In early 1913, however, Bohr became aware of work by the Swiss physicist Johann Jakob Balmer that fitted a simple formula to the spectral lines of hydrogen.

It was known that all elements emit and absorb specific frequencies of light. As far back as 1802, English scientist William Wollaston (the same Wollaston who had accused Michael Faraday of plagiarism) had discovered that sunlight did not contain a continuous rainbow spectrum, but rather had black lines where some colours were missing. These were frequencies where elements were absorbing light, just as elements would give off particular colours when heated. This mechanism was responsible for the discovery of helium, when unknown absorption lines were found in the solar spectrum (hence it being named after Helios, the ancient Greek sun god). Balmer had discovered a way to predict the position of these dark lines for hydrogen.

The existence of these specific, limited frequency absorptions and emissions filled in the last piece of the Manchester jigsaw for Bohr. It became clear to him that these fixed frequencies corresponded to limited orbits. Unlike a classical planetary orbit, which can take place at any height above the planet, the electrons in the atom could only exist in specific, discrete orbits, which he called stationary states. He arbitrarily decided that the electron would not be allowed to move to any orbit between the states, including the entire region between the lowest possible energy state and the nucleus, forcing stability on the atom. It was as if he put atomic electrons on rails, with a super power of jumping between them.

Bohr published his theory in three parts in the summer of 1913, once he was firmly established back in Denmark, but it is clear that the formulation of his atomic model began in Manchester, lacking only the insight that would emerge from his discovery of Balmer's theory. It was working at Manchester with the individuals he met there that proved the trigger for one of the most remarkable developments in all of physics.

As Bohr put it, it was at Manchester that he experienced 'the enthusiasm with which the new prospects for the whole of physics and chemical science, opened by the discovery of the atomic nucleus, were discussed in the spring of 1912 among the pupils of Rutherford'. By contrast with the seemingly cold, detached reception that Thomson provided, Bohr found the approach taken in Manchester much more conducive to developing his ideas. Every afternoon there was an opportunity to discuss new ideas over tea and cake, informal get-togethers often presided over by Rutherford, alongside his more formal Friday afternoon colloquia. The big difference for Bohr seems to have been this more collaborative, information-sharing approach. In Manchester, in the heady atmosphere of Rutherford's lab, the quantum atom was conceived.

Bohr's quantum model of the atom was at the heart of the first generation of quantum theory, and in 1922 Bohr deservedly received the Nobel Prize in Physics for this work. His model was successful in providing a stable model for the atom and the existence of atomic spectra. However, it did not explain the intensity of the different spectral lines and it was riddled with ambiguous behaviour and inconsistencies of interpretation. In the mid-1920s it would be transformed by quantum mechanics, where Heisenberg's matrix mechanics and Schrödinger's quantum-wave equation came together to provide a solid structure that would replace the hazy foundations of Bohr's pioneering model.

Bohr had discovered a way of looking at the atom that made the role of electrons clear and their position in (at least some) atoms stable, even if it was very different from anything we experience at the macro level of everyday life. It was at around the same time that the electron began to exert its power in a technology that was named after it: electronics.

STEAMPUNK ELECTRONICS 24

Where a calculator like the ENIAC today is equipped with 18,000 vacuum tubes and weighs 30 tons, computers in the future may have only 1,000 vacuum tubes and perhaps weigh only 1½ tons.

Andrew Hamilton in 'Brains that Click', *Popular Mechanics* (1949)

According to the Dutch-born physicist and science historian Abraham Pais, it was the tradition at annual Cavendish Laboratory dinners for some years after the electron's discovery to raise the toast, 'To the electron – may it never be of any use to anybody.'* In one sense, of course, this was hardly a realistic desire. Given that without the electron there could be no atomic structure, no chemistry and far less in the way of electrical phenomena – not to mention no living being to theorise about it – it had always been of essential use to everybody. But a more direct and hands-on use would come with the development of electronics.

Electron wrangling begins

We tend to think of electronics as electron-based engineering that involves solid state components, from transistors to integrated

* Pais's claim of the useless electron toast at the Cavendish was in *The Discovery of the Electron*, Beam Line (Spring 1997).

circuits. The sort of thing that's packed into a phone or a TV. But there was a first generation of electronics that had more in common with a Crookes tube than anything we might now associate with the term. It was the electronics of the steam age, perhaps the only genuine incarnation of the fantasy concept of steampunk.

The development of electronics marked a coming of age in the electron's biography. Up to this point it had only existed in its natural state – now it had grown up and had a job to fulfil. Nothing it had done up to this point required anyone to be aware of its existence. From now on, though, it would have fully fledged personality as an entity whose existence needed to be considered to get the most useful work out of it.

Before we reach the technology itself, it's worth spending a moment thinking about what we need electronic components to do. We usually think of electronic devices in terms of their application – a phone, a smart speaker, a computer, the control circuits of a car – but each of these applications depends on combining millions of individual components, the building blocks of electronic circuits.

Although there are many more complex possibilities involved, the vast majority of these components have the job of acting on electrons in one of six simple ways. Some can provide resistance – make it harder for electrons to flow. Other components act as one-way gates, only allowing electrons to pass in one direction. They can provide controllers to amplify a flow of electrons, or to act as an on/off switch for them. They can store electrons away for future use. They can convert the energy of electrons into another form. Or, the inverse of the last, they can convert another source of energy, such as light, into energetic electrons.

We have already met three of these in early forms. The light bulb converts the electron's energy into light (and heat). Resistance is a factor in anything electrons travel along (unless we're dealing with superconductors, which we met in chapter 22). And Leyden jars were the first useful storage mechanism for electrons. We'll come back to converting other energy sources into electron movement in chapter 32. But at the heart of making electronics in the modern sense – for example to act as an amplifier when playing music, or to perform basic computing functions – the essential components we need to add are one-way gates and controllers.

The Edison effect

The one-way gate was a classic example of a discovery being made before practical applications were possible. Several experimenters had noticed that a negatively charged piece of metal would somehow lose its charge into the atmosphere when heated. Edison, trying to understand the way his light bulbs lost material from the filament, discovered that there was a flow of electricity from a heated filament if he made a circuit inside the bulb with the heated wire on the negative side of the circuit and an additional, separate wire that was the positive side. This oddity, referred to as the Edison effect, had limited practical use initially.

However, the English physicist and engineer John Fleming, based at University College, London would find an unexpected implication of the effect. Unlike many of the scientists we have met so far, Fleming had something in common with Faraday and Edison in being from a poor background. Born in Lancaster in 1849 and moving with his family to London when he was four or five, the teenage Fleming had hoped to gain an apprenticeship in engineering, but his family couldn't afford the fees: instead he attended University College, London – though even here he was forced to stay away from the course for several months and only return on a part-time basis due to financial difficulties.

He briefly taught before gaining a DSc from Cambridge, staying for a short while at the university before rapidly moving through a chair in physics and maths at University College, Nottingham, three years working for Edison Lighting Company and finally arriving back at University College where he would remain until his retirement.

Fleming was involved in radio from the early days, designing the transmitter used by Marconi for his transatlantic tests (something Marconi singularly failed to reveal). In the early days of wireless transmission, detection of the radio signal involved a device called a coherer. This was a small vacuum tube with a pair of slightly separated electrodes, and some loose iron filings inside. Because of the gap between the electrodes, the device would not allow a current to flow – but a radio signal induced a charge that made the iron filings stick together, linking the electrodes and providing a conductor, enabling the signal to be detected.

Coherers were troublesome, inconsistent in their detection and required relatively strong signals before the effect worked. Fleming, aware of the Edison effect, wondered if it was possible to use a feature of the effect to turn high-frequency radio oscillations into useful signals. Although it wasn't known at the time, what was happening in the Edison effect was that the heated wire was giving off electrons, which were attracted across the vacuum to the positive electrode and allowing a current to flow. But this could only happen in the one direction. If the circuit was reversed so the heated wire was the positive side, no electrons would flow, because the negative electrode was not heated.

This meant a device using the effect only allowed a current to flow through it in one direction – it was that essential one-way gate – and in 1904 Fleming was able to use it to turn a high-frequency radio signal into easily detectible pulses. He called his device an oscillating valve, though it would become more widely known as the Fleming valve. The mental model of electricity was still very much that of a flow of liquid. This 'valve' only allowed a current of electricity to flow in one direction.

A valve for electric current

Such devices would be given the general name of thermionic valves, as they relied on a heated electrode – this particular type of valve was called a 'diode' because it only had two electrodes, the heated one and the cold one. The name would stick to electronic components with the same function long after thermionic valves were abandoned for solid state devices.* Not only did the diode help with radio receiver design, its one-way nature made it possible to 'rectify' AC current, converting it to DC – and in an electronic circuit it would prove essential to corral the flow of electrons in the required fashion.

* Thermionic valves are known as vacuum tubes in the US.

Four years later, the American electrical engineer Lee de Forest took the valve to the next level by adding a third electrode in the form of a grid between the heated wire and the cold wire. Born in Council Bluffs, Iowa, in 1873, de Forest is sometimes portrayed as if he were a backwoods tinkerer compared with other scientists and engineers working with electrons, but he had received a PhD from Yale University on electromagnetic waves. Admittedly, his experiments at the university regularly blew the fuses of the early electrical systems and caused blackouts, resulting in de Forest being thrown out of the University's Sheffield Scientific School. But after a brief involvement in the Spanish–American War of 1898, he returned to Yale, now working at the Sloane Physics Laboratory, to gain his doctorate.

After leaving university, like Fleming, de Forest worked on radio receivers, and later, as an independent inventor, he devised new mechanisms to improve radio transmission and reception. But his huge contribution to electronics would be the thermionic valve he would market as a 'grid Audion'. In 1906, he had started to work on diodes, which was when he introduced the Audion name – but it would be with his 1908 patent of the triode version that he struck gold (if not for himself).

Bear in mind that the Fleming valve worked because electrons were flowing from the heated negative wire to the positive electrode. De Forest realised that if he applied a negative charge to a grid positioned between the two electrodes, it would prevent electrons getting through. With a strong negative charge, this would act as a switch, turning off the flow when the charge was applied. This would make his grid Audion an equivalent of relays, mechanical switches that were already in use. Or if there were a small, varying value of negative charge applied to the grid, this would be reflected in a much stronger varying value of flow of current between the main electrodes. The triode would amplify a weak signal.

De Forest's triode provided another essential of electronic design. Like Tesla, he had a limited understanding of the mechanism that enabled his invention to work, believing that it was ionised gas in imperfectly evacuated tubes that was carrying the current – but the triode still did the job. De Forest would sell some of his patent rights to the American Telephone and Telegraph Company (AT&T) for $50,000 in 1913 – a not-inconsiderable sum of around $1.7 million in today's terms with inflation, or $9.6 million in terms of today's wages, but one that hardly recognised the importance of the invention.* His other patent rights would also later be bought by AT&T.

As we will discover in chapter 31, this ability to switch and amplify flows of electrons would eventually be replaced in solid state electronics by the transistor. For the moment, though, thermionic valves would transform the nascent radio industry, acting both to extract signals from a radio carrier wave and to amplify the weak incoming signal to provide effective listening. Valves would be common in household electronics all the way through to the 1960s, and remained in high-power commercial use for a decade or two more until solid state electronic components capable of managing high power signals were available. Some enthusiasts still continue to seek valves out because they believe them to produce a 'warmer' sound. There is some doubt that this preference would stand up to blind listening, but there is still something visceral about the glow of vacuum tubes in action.

Logically switching electrons

Where radio primarily relied on triode valves to act as amplifiers, from the 1940s it would be their role as switches to turn on and off the flow of electrons that would make them central to the new invention of electronic computers. Computer logic is dependent on having objects that can be in binary states – on or off, for example. This could be achieved using electromechanical relays, but they were prone to sticking and difficult to use at any scale. But triode valves would enable the construction of more effective computing devices.

Within computers, electronic components are used to build next-level structures known as gates. These most frequently work on one or two values at a time and perform logical functions. So, for instance, a NOT gate provides no voltage if there is a voltage input,

---

* Much as there had been between Edison and Swan over the light bulb, there were several patent disputes over thermionic valves, both between Fleming and de Forest and with other US companies over alleged patent infringements.

but outputs a voltage if no voltage is fed in. Thinking of this in terms of bits, it converts a 1 into a 0 and a 0 into a 1. A single triode, with some help from a couple of resistors, can do the job of a NOT gate, with the input voltage being applied to the grid, so it switches the voltage across the other two electrodes on and off.

Valves were not without problems in practical use. Like incandescent light bulbs, they wasted a lot of energy as heat, they burned out relatively quickly and were fragile glass structures that were easily broken. But with the single-application Colossus code-breaking electronic computers used at Britain's Bletchley Park in the final years of the Second World War, and most particularly with the development of ENIAC in the US at the end of 1945, they would prove their worth.

ENIAC (Electronic Numerical Integrator and Computer), initially housed at the University of Pennsylvania, is generally recognised as the first general purpose electronic computer. It incorporated over 17,000 valves, which had to be run at so high a voltage that they failed frequently. The typical runtime of ENIAC before a failure was two days, and it never kept going for more than five days before engineers had to make the search for the burned-out component causing a failure. Not surprisingly, this could take some time.

ENIAC weighed around 27 tonnes and needed around 150 kilowatts of electric power to run, occupying a room around 30 metres (100 feet) long.* It is difficult to make direct comparisons with modern devices given the different structures involved, but the scientific processing power of computers is often measured in FLOPs (floating-point operations per second). A high-end smartphone might easily achieve 5 trillion FLOPs – ENIAC achieved about 500. The forecast at the opening of this chapter that computers might one day weigh as little

* The impractical nature of early computers explains why Thomas Watson Sr., the head of IBM in the 1940s, is alleged to have made the disastrous prediction that the world probably only had a demand for about five computers. It has been suggested this was a misattribution based on Howard Aiken's, one of IBM's lead engineers, 1952 remark, 'Originally one thought that if there were a half dozen large computers in this country, hidden away in research laboratories, this would take care of all requirements we had throughout the country.'

as 1½ tons has been somewhat exceeded with devices running billions of times faster.

As a result of the introduction of electronics, the free electron had a new, more subtle practical role rather than being simply the constituent of electricity. But there was also an increasing understanding of the electron's vital role in atoms. Electrons are responsible for most of the physical interactions that atoms undergo, shielding the atomic nucleus from the outside world.

Think, for instance, when you sit on a chair. Neither your body nor the chair itself is a continuous solid. Atoms are mostly empty space,* with the massive nucleus occupying a tiny fraction of their volume. There is no solid connection between atoms – instead it is the electromagnetic interaction of the components of the atoms that holds them together. And it is only the repulsion between the electrons on the outside of the chair's atoms and your own that prevents you passing straight through the seat.

This is just the start of the electron's fundamental role in the structure of the atom. As we will discover, the electron rules over chemistry and biology. Each of these major scientific fields are dependent on the electrons that orbit atoms.

---

* It is popular among scientific killjoys to say that the statement that atoms are mostly empty space is a science communication myth. This is because the quantum models of the atom we will discover soon say that electrons are not located in a discrete position at any one moment, but inhabit a fuzzy region of probability, filling quite a lot of the internal space of the atom. But this itself is based on something of a myth, that quantum particles can be in more than one place at a time. In fact, a better description would be that they aren't anywhere until they interact with something else: all that exists are the probabilities. Given that definition, it's still reasonable to say that atoms are mostly empty space.

25 ORBITALS AND ATOMS

The atom of the chemist is now a reality; but that does not mean that we are close to touching the ultimate elements of things. When Democritus invented atoms, he considered them as absolutely indivisible elements and beyond which there is nothing to look for. That's what it means in Greek; and that is why he invented them; behind the atom, he did not want any more mystery. The atom of the chemist would not have given him satisfaction, for this atom is by no means indivisible, it is not a true element, it is not without mystery; this atom is a world … in the atom we find many other things: first we find electrons; each atom then appears to us as a kind of solar system, where small negative electrons acting as planets revolve around a large positive electron that plays the role of the central sun. It is the mutual attraction of these electricities of contrary name which maintains the cohesion of the system and which makes it a whole; it governs the periods of the planets, and it is these periods which determine the wavelength of the light emitted by the atom; it is to the self-induction of the currents of convection produced by the movements of these electrons that the atom which is formed owes its apparent inertia and which we call its mass. Besides these captive electrons, there are free electrons, those which obey the same kinetic laws as the gaseous molecules, those which render the metals conductive. These are comparable to comets that circulate from one star system to another and establish between these distant systems as a free exchange of energy.

Henri Poincaré in *Reports on Matter and Ether* presented at the French Society of Physics (1912)

When Henri Poincaré wrote about the structure of the atom and its implications for chemistry, Niels Bohr was yet to present his initial quantum model of the atom, meaning that Poincaré could still put forward a solar system model for atomic structure. As we will see, this would rapidly be replaced by a quantum structure where the locations of electrons were far from certain. By 1912, though, we can consider the electron to have been brought to maturity by physicists – and it would prove to be entirely central to chemistry.* J.J. Thomson was arguably the first to really drive the idea that chemical interactions were primarily electronic, relying on the interaction of corpuscles (electrons) on the most external layer of the atom.

The key to chemistry

Anyone who has studied basic chemistry will be aware of the concept of chemical bonds – the links between atoms that produce molecules. Initially, Thomson only considered bonds involving electrical charges. So, for instance, common salt, sodium chloride, is the result of the positively charged sodium ion being electrically attracted to the negatively charged chlorine ion, the sodium ion being short of an electron and the chlorine ion having one extra.

Thomson also related the outer electrical structure of the atom to the periods of the periodic table (see chapter 10), where elements were grouped by the number of electrons or gaps in their outermost electron orbit. These 'polar' ionic bonds, though, would prove insufficient to explain all molecules, particularly the majority of organic compounds where bonds between, for example, carbon and hydrogen had an electrical symmetry that did not seem to reflect the link of a positively and negatively charged ion.

Inspired in part by Thomson's observations, American chemist Gilbert Lewis would realise that in this non-polar type of bond, rather than, say, an electron belonging to the chloride ion ending up taking part in the outer shell of a sodium ion, each of the elements would

* There's considerable irony to saying that the electron is *central* to chemistry, as its importance there is entirely due to electrons being on the outskirts of the atom.

effectively be contributing an electron to share between the two of them in the bond, which would become known as covalent.

When these ideas were coming together, Bohr's model of the atom was already known, but arguably because the tools weren't available to properly deal with such a structure where the outer part of the atom is in a constant state of change, the early ideas of chemical bonding were primarily developed thinking of the atom as a static structure, more like a crystal than a system featuring constant movement and, indeed, uncertainty of location.

Predicting quantum behaviour

As quantum physics was established in the 1920s and 30s, it presented a very different view of electrons to those held by earlier physicists. It was natural to think of electrons as particles just as we might think of, say, a particle of dust – just much smaller. But as physicists began to develop the theory needed to explain the behaviour of quantum particles, the picture became far stranger. The German physicist Werner Heisenberg felt the need to throw away Bohr's picture of orbits, which took as their model the visible world, opting instead for a purely mathematical view, based on matrices (two-dimensional tables of numbers with a whole new variety of arithmetic to support them), which accurately predicted the behaviour of quantum particles without any picture of what was happening.

Heisenberg's 'matrix mechanics' was important, but many physicists were more comfortable with an alternative approach developed by Austrian physicist Erwin Schrödinger. The two were entirely equivalent, but Schrödinger made use of a model that seemed to offer better parallels with phenomena we can directly experience. Born in Vienna in 1887 and by 1920 an associate professor at the University of Stuttgart (arguably this would have been earlier had it not been for his time in the army during the First World War), Schrödinger was without doubt a brilliant physicist, though he was always unconventional in his domestic arrangements, living with his wife and another woman for decades, and accused of a predilection for underage girls.

For Schrödinger, the ability to visualise what was happening was essential: he devised an alternative description of quantum physics based on waves. Unfortunately, there were two significant problems with this approach. Firstly, Schrödinger's wave equation seemed to suggest that over time quantum particles should spread out over more and more space. And secondly, the initial solutions to his equations contained imaginary numbers.

Imaginary numbers are based on the square root of -1. The concept was considered meaningless for many years because in conventional arithmetic -1 has no square root. Since the square root is the value that when multiplied by itself produces the desired outcome, this presents a problem. Multiplying positive numbers by themselves obviously produces a positive number – but so does multiplying a negative number by itself.

Imaginary numbers were first introduced by Italian mathematician Gerolamo Cardano in the mid-1500s, though he considered them without practical value. The square root of -1 was given the identifier *i* in the eighteenth century, which became useful as it was around the 1730s when it was realised that Cardano was wrong. Imaginary numbers could be considered a marker for a second axis at right angles to a first. By combining a real number and an imaginary number in what became known as complex numbers such as $3+4i$, mathematicians discovered they now had a tool that was extremely useful for anything involving changes of values in two dimensions, such as waves.

Physicists and engineers began to use complex numbers regularly – but always in such a way that there was never an imaginary number left in their final result. Wave mechanics, as Schrödinger's approach to quantum theory was described, did not keep things real. The conundrum that Schrödinger had uncovered was cracked by German-British physicist Max Born, a close friend of Einstein's. He realised that Schrödinger's equations described not the position of a particle, but the probability of finding it at a particular location – at least, when the outcome of the equation was squared.* And by squaring it, those irritating imaginary values disappeared.

* Born's introduction of probability to the location of particles was what inspired some of Einstein's best-known quotes, mostly in letters written to Born. These included the idea that God does not play dice. In one letter he said if it were true, 'In that case, I would rather be a cobbler or even an employee in a gaming house, than a physicist.' Although Einstein's explanation of the photoelectric effect was a major part of the foundations of quantum physics, he was never reconciled to the theory's implications.

In one sense this probabilistic approach to a particle's location seems reasonable. If we imagine, for instance, marking the position of a horse in a field, then look away for half an hour, we could not know where the horse was. But we could say there was a higher probability of finding it in some locations (perhaps by a water trough) than in others – such as in a patch of bushes, or up a tree. However, there is a fundamental disparity between these two pictures. The basis of using probability for the horse's location is a lack of knowledge. The horse will be in one specific location, but without knowing what it is we can assign probabilities to where it might be. With the quantum particle, things are entirely different.

For an electron, for instance, quantum theory tells us that, before we observe it or it interacts with something else, the electron cannot be located. All that exists is the collection of probabilities of finding it in various places. The electron is not anywhere specific. It's important to note that the probabilities involved are entirely predictable – they can be calculated with great accuracy. But until the electron undergoes an interaction with the outside world, we cannot say where it will be discovered.

Bringing quantum reality to the atom

Back in the atom, the understanding of how electrons were distributed finally provided an explanation for the strange asymmetric structure of the periodic table, uncovered more and more after Mendeleev's original work we saw in chapter 10. One way of looking at the electronic structure of an atom is to consider the electrons to be in 'shells' – these correspond to the different energy levels electrons are allowed to have, with each shell potentially containing a number of 'orbitals'.

As we have seen, these are not orbits in the same sense of the orbit of the Earth around the Sun. Astronomical orbits are like our horse – the Earth is always somewhere specific. In fact, the orbit is much more predictable than the horse, following simple equations of motion under gravity. By contrast, the orbitals inhabited by electrons in an atom are probability distributions predicted by Schrödinger's equation – three-dimensional shapes in which a particular electron may be found over time, with differing probabilities.

The position of an element in the periodic table depends on how different shells are filled with electrons. From the chemist's viewpoint, it is only the outermost shell that contributes to an element's reactions. This is the one that will establish its chemical properties. The way that shells fill is perhaps not one we would design by choice – all we can do is follow an empirical rule of what actually happens. The first shell surrounding the nucleus can only hold two electrons, resulting in the isolation of hydrogen (with a single electron) and helium (with two) at the top of the periodic table.

Chemical reactions are most easily made by atoms with an incomplete outer shell. So hydrogen, with a space free in its only shell, is far more reactive than helium with the filled shell. The same applies to all the so-called noble gases at the right-most side of the periodic table – neon, argon, krypton and so on joining helium. Each has a filled outer shell and rarely reacts chemically.

Strangely filling shells

Where things get particularly complicated, resulting in the odd shape of the complete periodic table, is that after the first shell, the outermost shell can only hold a maximum of eight electrons, even though it may have a far greater capacity. So, for example, the third shell can hold eighteen electrons, but as the atom gets bigger and more electrons are present, when the third shell has eight in place, the fourth shell will start to fill. Now, though, the third shell is no longer the outer shell, so it can start to hold extra electrons, all the way up to that total of eighteen.

The result is that the periods (rows) of the periodic table have a messy, staggered structure. The first has two elements in it, the second and third eight elements, the fourth and fifth 18 elements and the sixth and seventh 32. While the exact derivation is somewhat messy, the limitations of each shell are driven by a combination of subshell capacity and the Pauli exclusion principle.

The exclusion principle is a fundamental rule of quantum physics. It says that within a quantum system, it is impossible for a class of particles called fermions to be in the same quantum state. This 'state' sums a number of properties of the particle, known as its quantum numbers – more on those in a moment. Fermions are particles with one of its properties, quantum spin, holding half-integer values – usually ½.* These are the electron, proton, neutron, neutrino and quarks.† (Other quantum particles, known as bosons, including photons, have integer spin and can crowd together as much as you like.)

Electron spin only has two possible values, ½ and -½, leaving room for two electrons in the first shell, as all the other quantum numbers so close to the nucleus remain the same. As we move outward to extra shells, other quantum numbers come into play. Each shell has a different principal quantum number: that immediately gives us the ability to have two electrons in the second shell. But they can also vary by the azimuthal quantum number, which is effectively based on the shape of orbitals. The breakdown of orbitals by azimuthal quantum number is described as subshells, and the azimuthal number can have the same number of values as the shell number.

This means the second shell has two subshells, each capable of housing electrons of two different spin values. But there is also the impact of the magnetic quantum number, which in effect says how the orbitals are oriented. This has values that increase with the number of azimuthal quantum number values, resulting collectively in the odd-feeling increase of number of electrons possible in each shell.

Electrons and the way they can fit into an atom thanks to quantum limitations are central both to the structure of atoms and, down to the outermost shell, determining the chemical nature of each element. Chemistry is a purely electronic science. In many ways, this is the case for biology too. Although a good number of biological electro-chemical mechanisms are related to the movement of protons, most biological

---

* Although 'spin' sounds as if it refers to a particle spinning around, it is a quantised property that has no correspondence in physical objects we can directly experience.

† We are generally happy that, like, for instance, Einstein or Sting, the electron can get away with a single name, but including its family name presumably makes it electron fermion. Electrons belong to a sub-family of fermions called leptons, so perhaps in full that should be electron lepton fermion.

realities are driven by the chemistry of the often-complex molecules involved.

Biological oceanographer Paul Falkowski has referred dramatically to the 'electron market' in living organisms. He points out that nature operates at the level of nanomachines, responsible for processing molecules, such as oxygen and carbon dioxide – and the operation of these invisibly small machines is electrical, based as it is on chemistry which is driven by the interaction of electrons. He sees the natural environment – rivers, oceans, soil, atmosphere – acting effectively as the wires that enable electronic action to occur, transferring electrons between living organisms (mostly, numerically, single-celled) with this electron market being the underlying mechanism for life.

The new understanding of electrons as quantum particles would be extended by the realisation that, just like light, it was possible for the components of matter to be regarded as particles or waves. It was time for electrons to go wavy.

WAVY ELECTRONS

26

Of course, if electrons were waves there would be no difficulty. We think we understand the regular reflection of light and of x-rays – and we should understand the reflection of electrons as well if electrons were only waves instead of particles. This observation though true does not seem a particularly valuable one. It is rather as if one were to see a rabbit climbing a tree, and were to say, 'Well, that is rather a strange thing for a rabbit to be doing, but after all there is really nothing to get excited about. Cats climb trees – so that, if the rabbit were only a cat, we would understand its behavior perfectly.'

Clinton Davisson in 'Are Electrons Waves',
Journal of the Franklin Institute (1926)

Just as is the case with political dynasties, we occasionally see a family where more than one generation makes an impact on our scientific understanding. Two families are probably best known for this by the public: the UK has the assorted Darwins who have graced science (most notably after Charles were astronomer George and the largely twentieth-century physicist, also, confusingly, Charles). Meanwhile, in France, Marie Curie and her daughter Irène were the only mother and daughter to have each won a Nobel Prize. However, in the story of the electron, it's a father and son – also both Nobel Prize winners – that get the accolade.

Just as J.J. Thomson had established the existence of the electron,* 30 years later his son George demonstrated that electrons could act like waves using the technique of electron diffraction. This would be one of the fundamental steps forward in developing quantum theory, also bringing in the work of a French nobleman. Just as light waves could act as the particles known as photons, so electron particles could behave as if they were waves. Suddenly, the central character in the story of electricity had become more mysterious still.

Dr Thomson Junior and the Duc

George Thomson had been brought up in and out of the Cavendish Laboratory, soaking up maths and physics from an early age. After postgraduate work at the Cavendish, in 1922, aged 30, he was appointed Professor of Natural Philosophy at the University of Aberdeen, an inverted journey to James Clerk Maxwell, who had been a professor in Aberdeen before setting up the Cavendish in Cambridge. It took George some years to get an old-fashioned and somewhat neglected laboratory up to speed, during which time a new philosophy of the nature of quantum entities had emerged on the continent. It would come from no less than a French duke.

Technically, Louis de Broglie was the 'Seventh Duc de Broglie'. We rarely think of aristocrats as potential scientists, but Louis' older brother Maurice (the Sixth Duc) was an experimental physicist, and though Louis' first degree was in history, he soon succumbed to the siren song of physics. Born in Dieppe in 1892, de Broglie already had a second degree in physics under his belt when the First World War broke out, leading to work on military radio, particularly communicating with submarines.

By 1924, inspired by the way that light appeared to act both as if it were a particle and a wave, de Broglie applied the same thinking to the electron. Without experimental verification he predicted that

* Technically, J.J. Thomson's Nobel Prize was not for the discovery of the electron, but for his work on the electrical conductivity of gases, but the two were tightly tied together.

the wavelength of a particle would be inversely proportional to its momentum (i.e. if its mass times velocity grew, its wavelength would decrease), while its frequency would be related to its energy.

In his 1924 doctoral thesis, de Broglie noted that he was 'inclined to admit that any moving body may be accompanied by a wave and it is impossible to disjoin motion of body and propagation of wave'. Arguably, this implies that even, say, a human would have an associated wave, though they would have to be moving incredibly slowly for their momentum to be sufficiently low for the wavelength to be potentially measurable.

On the theoretical side of physics, this would have a profound effect. The best mathematical description of quantum physics at the time was matrix mechanics, which, as we have seen, relied on very abstract mathematical structures to make calculations. Wave mechanics, devised by Erwin Schrödinger, would make equivalent calculations based on the more familiar structure of a wave equation, problematic though that may have been.

However, it wasn't until 1927 that George Thomson in Aberdeen, working with his graduate student Alexander Reid, pretty much simultaneously with Clinton Davisson and Lester Germer at Bell Labs in America, was able to demonstrate electrons acting like a wave. For George it was a parting of the ways with his father. When he first heard of de Broglie's theory, he attempted to merge it with his father's dedication to Faraday tubes and the aether – but by 1926 he had moved on to search for diffraction, a wave-like behaviour, in particulate electrons.

Diffraction and interference

Diffraction involves a wave changing direction when going through a narrow gap or hitting an obstacle. A simple type of diffraction is when a wave passes through a narrow gap. Instead of heading straight forwards on the other side of the gap, the wave opens out from the gap, heading off in all available directions as if it were a new wave starting from the gap. This kind of diffraction takes place in the famous Young's slits experiment, where a beam of light passes through two slits and the resultant waves from the slits interfere with each other to cause a pattern.

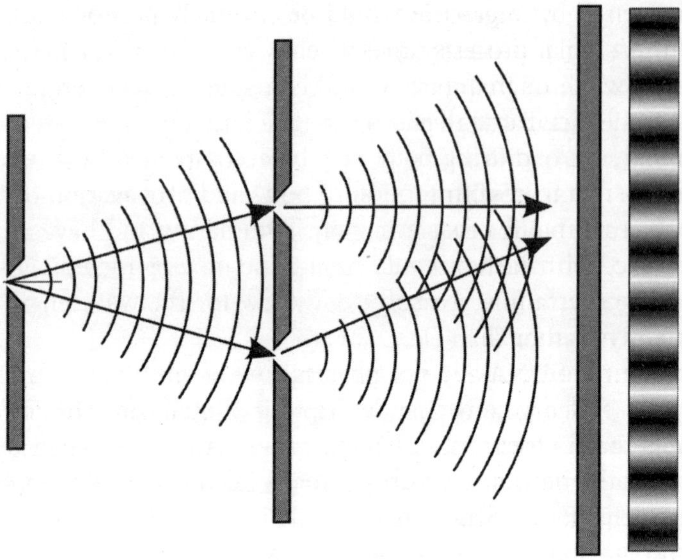

Young's slits experiment.

It would later be demonstrated that this same kind of dual slit experiment could be performed using electrons. Moving to electrons could then be used to show just how strange quantum behaviour is. Though it has since been demonstrated with photons too, using electrons made it possible first to send particles through the experiment one at a time. In classical physics, this should produce just two blobs on a detector the other side of the slits, corresponding to whichever slit each electron passed through.

In reality, however, even when firing one particle at a time, an interference pattern of dark and light fringes builds up. It's popular to say that this is because the quantum particle (whether it be electron or photon) passes through both slits at once and interferes with itself. But that misses the reality of quantum particles existing as a cloud of probability until, in this case, they interact with the detector. It's an interaction of probability waves, not a particle passing through both slits.

When Thomson (and Davisson) showed the first example of electron diffraction, though, they were not using twin slits. You see a diffraction effect when a CD or DVD produces a rainbow pattern, as the pits in the surface of the disc that store the data act as a diffraction grating where different colours of light are shifted to a different degree.

This is caused by interaction of the probability waves of the photons of light. A similar process happens with electrons when, for example, they are sent through a thin film of gold, nickel and other metals: the atoms in the metal act as a grating on the scale of electron wavelengths.

In this kind of diffraction, electrons fired through the crystal structure of the metal are diffracted as they interact with the metal atoms, with the result being new sources of waves at each interaction causing interference with each other and generating a pattern that is dependent on the crystalline structure. It would be a similar process, but using X-rays rather than electrons, that would be used to determine the structure of DNA as a result of the diffraction patterns caused.

By 1930, George was able to acknowledge the parting of the ways. Although both electrons and photons could act as waves, one could be electrically deflected and the other couldn't. Had they both been waves in the aether, this would mean that there had to be (at least) two different types of aether, which seemed an unnecessary complication for George, who was able to remove the mental crutch of the aether that his father would never fully relinquish.

George shared the Nobel Prize in Physics for 1937 with Clinton Davisson. In his Nobel Lecture, George noted, 'I should be sorry to leave you with the impression that electron diffraction was of interest only to those concerned with the fundamentals of physics. It has important practical applications to the study of surface effects. You know how X-ray diffraction has made it possible to determine the arrangement of the atoms in a great variety of solids and even liquids. X-rays are very penetrating, and any structure peculiar to the surface of a body will be likely to be overlooked, for its effect is swamped in that of the much greater mass of underlying material. Electrons only affect layers of a few atoms, or at most tens of atoms, in thickness, and so are eminently suited for the purpose. The position of the beams diffracted from a surface enables us, at least in many cases, to determine the arrangement of the atoms in the surface.'

When we look at an object, we are, in effect, determining the arrangement of atoms in the surface. It's what we do when looking through a microscope. But optical microscopes are limited by the wavelength of light – electrons, with their much smaller wavelengths, had the potential to resolve much greater detail. What George was noting here was the emerging concept of an electron microscope.

Seeing with electrons

Although early attempts to employ what could be considered a microscope principle were made by detecting a metal grid in 1931, it wasn't until 1932 that German physicist Ernst Ruska (who won the 1986 Nobel Prize in Physics for his work on electron 'optics') suggested that de Broglie's electron waves meant that the electron could provide an alternative to light for microscopic work. Working at the German technology firm Siemens, in 1939 he and his colleagues produced the first commercial electron microscope.

Since then, a range of electron microscope types have been developed. The initial devices were transmission electron microscopes, which required a thin sample so that electrons could pass through it and be detected the other side. Perhaps the most familiar, though, in terms of images produced, is the scanning electron microscope (SEM), which doesn't require a thin 'transparent(ish) to electrons' sample, but instead gives a textured image of the surface of a sample.

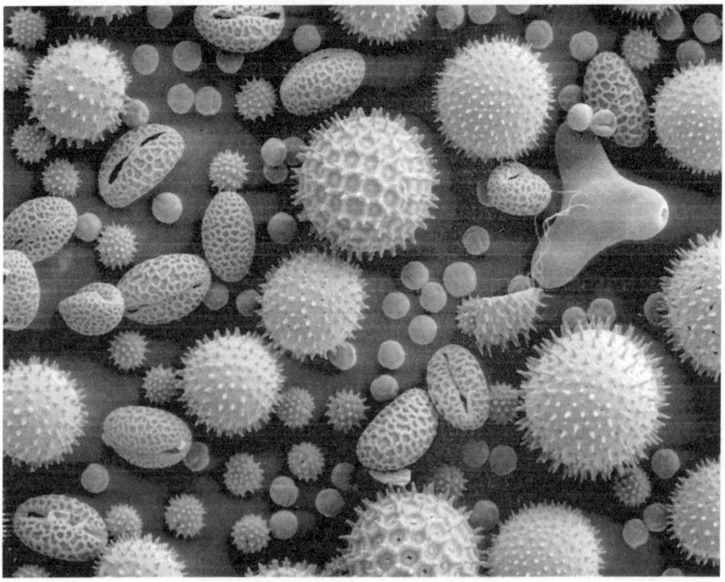

SEM image of pollen grains.

Dartmouth College Electron Microscope Facility

Public domain, via Wikimedia Commons

In an SEM, the electron beam stimulates electrons in the surface of the material studied sufficiently for them to be emitted back towards a detector which uses these secondary electrons to build a picture of the surface. Perhaps the other best-known approach isn't an electron microscope in the same sense, though it is a microscope using electrons. This is the scanning tunnelling microscope (STM).

Rather than have an approach analogous to an optical microscope, an STM hovers a tiny, pointed probe close to the surface of the material being scanned. Electrons from the probe jump* across the vacuum that is between the probe and the surface using the quantum process known as tunnelling (something we will examine in more detail in chapter 31). A picture of the surface which can resolve individual atoms is built up by monitoring the tunnelling current. The microscope can also be used to manipulate individual atoms, most famously when IBM spelled out their initials using individual xenon atoms.

The majority of electrons we are likely to use in such microscopes were formed in the very early moments of the existence of the universe and have been around ever since. But we have also discovered that some electrons are born from processes in the atomic nucleus, where no electrons exist, but they can nonetheless be emitted. It's time to take a look at these newly birthed particles.

* Strictly speaking, in quantum tunnelling electrons do not jump across a gap, but rather emerge on the other side of the gap without having travelling through the space in between, but the English language is not entirely suited to the description of quantum processes.

NEW AND SHINY

27

There is a form of radioactivity that happens very slowly – the kind people worry about leaking out of nuclear reactors – called beta decay, which involves a neutron turning into a proton.

Richard Feynman in *QED: The Strange Theory of Light and Matter* (1985)

As we discovered in the first chapter, most of the electrons we encounter have been around for almost the entire lifetime of the universe. But some haven't. They are being created every day. These newborn electrons were discovered as a result of nuclear reactions in the process known as beta decay.

Exploring the nucleus

We need to backtrack a little to the gradual uncovering of the nature of the nucleus of an atom. We have seen how Thomson's plum pudding model was pushed aside by Rutherford's nuclear discoveries. The nucleus, much smaller in scale than the overall atom, appeared to be positively charged, and to contain most of the mass of the atom. In 1919, Rutherford had managed to winkle particles he would call protons out of nuclei by bombarding them with alpha particles (helium nuclei).

This slightly odd-sounding word, proton, came from the ancient Greek word meaning first, and Rutherford initially applied it to the nucleus of hydrogen, which contains a single proton, in 1920. This was the outcome of bombarding nitrogen with alpha particles, sometimes knocking out a positively charged particle that had (almost) the same mass as a hydrogen atom. This was the particle that Robert Millikan insisted in his Nobel Lecture should be called a 'positive electron' (see page 165). But his term was unnecessarily confusing.

The proton may have the same size electrical charge as the electron, if positive rather than negative, but it is a very different type of particle. Not only is it over 1,800 times more massive than an electron, we now know that it is not a fundamental particle, but is made up of three quarks. A far more appropriate target for the term 'positive electron' would be the positron (a much snappier name), which we will discover in the next chapter.

However, in itself, the discovery of the proton was not sufficient to explain the makeup of the nucleus. In most atoms, the nucleus appeared to be heavier than the appropriate number of protons needed to balance out the electrical charge of the electrons, producing a neutral atom. One possible solution to this quandary was that the nucleus contained electrons as well as protons – that not all the electrons in an atom inhabited those fuzzy external orbitals. In a sense this feels like a hangover from the plum pudding model. There, the atom had electrons embedded in the positive charge. Here, there would be electrons cosying up to at least some of the protons in the nucleus, so there could be more of them than was suggested by the number of external electrons.

This idea of nuclear electrons also fitted well with two of the types of radiation that Rutherford had originally given the labels of alpha, beta and gamma rays. The first of these had turned out to be helium nuclei, so became 'alpha particles' rather than rays. The second would turn out to be high-energy electrons, leaving only gamma rays as true electromagnetic radiation, high up the same spectrum as light. It seemed entirely reasonable that radioactive substances demonstrated the existence of both alpha particles and beta particles in the nucleus, combined with standalone protons to make up the required positive charge.

In 1932, English physicist James Chadwick, working at the Cavendish Laboratory in Cambridge, isolated a new particle with a slightly larger mass than a proton – the electrically neutral neutron.

Chadwick was born in Bollington, Cheshire, and hoped to attend my old school, Manchester Grammar School, but despite gaining a scholarship, his family could not afford to send him, so instead he attended the Manchester Central Grammar School.* He went on to Victoria University of Manchester in 1908 the year after Rutherford took over the physics department.

After graduating and gaining an MSc in 1912, Chadwick moved to part of the Physikalisch-Technische Bundesanstalt (the German equivalent of the UK's National Physical Laboratory), located in Berlin, working on beta radiation under one of Rutherford's former researchers, Hans Geiger. His timing was not ideal. Although he produced some interesting results, the outbreak of the First World War in 1914 meant that he was interned as an enemy alien, though he was able to conduct some ad-hoc research, including experimenting with radioactive toothpaste.†

After the war, Chadwick returned to Manchester, but in 1919 followed Rutherford to Cambridge to undertake his PhD there. He would go on to study the nucleus and beta radiation, particularly attempting to deal with technical problems that were thrown up by the idea of nuclear electrons. His would be the last of three research efforts attempting to clarify just what was happening in the nucleus.

A three-way struggle

The best known of the three key nuclear teams at the time was Irène Joliot-Curie and her husband Frédéric, working in Paris. They were looking into radiation produced when alpha particles (emitted by polonium, discovered by Irène's parents) were fired at relatively light

* Oddly, the Nobel website names his school as Manchester High School, which was and still is a girls' school.

† In these early years after the discovery of radioactivity, there were many products that used a radioactive label to suggest that their product was energising. Thankfully, many such products were not actually radioactive, but some, including a whitening toothpaste, and radioactive soda syphon cartridges genuinely were.

elements such as beryllium. This new radiation could not be deflected by an electric field. Of Rutherford's trio of rays, this suggested gamma rays were being produced. In early 1932, the husband-and-wife team used the impact of this radiation to produce high-energy protons from paraffin wax. But this would require far more energy than gamma rays could feasibly have provided.

Enter the second player in the field, the young Italian physicist Ettore Majorana. He has been described as the greatest physicist never to fulfil his potential. Majorana disappeared in 1938 aged just 32. Never totally explained, his disappearance has been blamed on everything from suicide to retirement to a monastery, or even murder to prevent his involvement in constructing a nuclear bomb. The most likely explanation seems to be that he emigrated to Venezuela and dropped out of science entirely, though there is no good evidence as to why this occurred.

Other physicists, notably Enrico Fermi, have suggested that Majorana could have been another Einstein or Newton, had he had the time to continue with his career. Majorana, studying the Joliot-Curies' results, suggested that the most likely possibility for this radiation that could smash out protons with such energy was a particle with a similar mass to the proton but with no electrical charge. Fermi wanted him to publish this result, but Majorana declined. Allegedly, this was because he believed that everything that could be understood was banal, and so did not need publication.

This left the field open for Chadwick. With assistance from the younger English physicist Norman Feather, he produced experimental support for the inability of gamma rays to be causing this disruption of nuclei. The evidence built to a level where Chadwick could first publish in February 1932 a paper entitled 'The Possible Existence of a Neutron', and then in May back it up with the definitive 'The Existence of a Neutron'. The name he chose for his new particle needs little explanation. Chadwick would be the sole winner of the 1935 Nobel Prize in physics for this work.

Where did the nuclear electrons come from?

Neutrons made for a simple picture of nuclei composed of just protons and neutrons, which was the standard description of the nucleus until the quark constituents of these two particle types were discovered. But this left the emergence of beta particle electrons as something of a puzzle. Chadwick initially assumed that a neutron was a compound particle, made up of a proton and electron, something that had been predicted as early as 1920 by Rutherford.

A compound particle would provide both electrical neutrality and the ability to apparently generate a new electron as a beta particle, but initially there was no suggestion that neutrons could decay, suggesting instead that beta particles were still separate parts of the nuclear mix. By 1934, though, Chadwick had become confident that the neutron was a new standalone particle. For some time, the mass of the neutron had not been measured accurately.

Perhaps counterintuitively, if it were less than the combined mass of a proton and electron, this suggested a compound particle. But if it were higher than the combined mass, it would be fundamental. This is because a component of the mass of a compound particle is tied up in its binding energy, the energy in nuclear bonds. When nuclear particles bind together, energy is given off which, thanks to $E = mc^2$, results in a reduction in mass. Chadwick, along with his graduate student Maurice Goldhaber, was able to make the first accurate measurement of neutron mass, which turned out to be slightly larger than the combination of proton and electron masses.

Just two years later, the relationship between nuclear particles and beta radiation would start to be explained, in part due to the theoretical prediction of the existence of yet another neutral particle – the neutrino. The existence of such a particle had been suggested a few years before in 1930 by Wolfgang Pauli, because at first sight, beta decay seemed to break conservation laws.

There are a whole set of properties, notably charge, energy, momentum and spin, that need to be conserved when quantum particles interact. While production of beta radiation was fine from a viewpoint of conservation of charge, something was missing when it came to the other measures. Niels Bohr would suggest this was a special case that broke conservation laws, but others were doubtful.

Pauli decided instead that there was an unknown particle involved in beta radiation, which he called a neutron. (Bear in mind this occurred two years before Chadwick came up with the name for the particle he discovered.) Pauli believed *his* neutron should have a mass similar to that of an electron – he imagined a kind of neutral electron. When Fermi was speaking of Pauli's hypothetical particle in July 1932, after Chadwick's discovery, he instead referred to it as a neutrino (using the Italian diminutive to suggest this was a little neutral particle), a name apparently suggested to him as a joke by Italian physicist Edoardo Amaldi.

It took a few more years to provide experimental evidence, but it became clear that neutrinos (or, to be more precise, antineutrinos, their antimatter equivalent) played a kind of midwife role in the production of electrons as beta radiation. A neutron was decaying in the nucleus to form a proton, which stayed put; an electron, which shot out as beta radiation; and an antineutrino, which also headed out of the nucleus but was extremely difficult to find. Rather than detect a neutrino, the experiments showed that the movement of the electron plus the recoil of the nucleus required the existence of this extra particle.

The full reality of what was happening in beta decay would only become clear once the compound nature of protons and neutrons was discovered. We now know that both of these particles are made up of three fundamental particles called quarks. Quarks come in six different 'flavours', but only two are required here. A proton is made up of two up quarks and one down quark, while a neutron is two down quarks and one up quark.

When a neutron undergoes beta decay, one of the neutron's down quarks changes into an up quark. This is a result of one of the four fundamental forces of nature called the weak nuclear force or weak interaction. Fundamental forces are transferred by so-called carrier particles. In this case, the carrier is the W^- boson. Unlike other forces, the weak interaction carriers are charged particles with mass: this negatively charged particle immediately decays into an electron and antineutrino. Thankfully, in a stable atomic nucleus, neutrons do not decay in this fashion, because it requires an input of energy to reach the combined energy state of the resultant particles.

It was with the realisation of the birth process of beta particles that new generations of electrons entered our understanding. The bulk of electrons remain those that have been around since shortly after the Big Bang. But whenever beta radiation has been produced, new shiny electrons have been added to the particle count. While the next electron you encounter is likely to be ancient, a few will be younger even than you are.

The explanation of beta radiation was tied into the development of quantum physics, as was the realisation that electrons could have an evil twin – one with an opposite charge that would be its nemesis. Conceptually devised to solve a problem with electrons having negative energy, this alternate-world electron was first predicted theoretically and then observed at the same time as the discovery of the neutron with its implications for electron birth.

THE ANTI-ELECTRON EMERGES 28

> *An unoccupied negative-energy state, or hole, as we may call it for brevity, will have a positive energy, since it is a place where there is a shortage of negative energy. A hole is, in fact, just like an ordinary particle, and its identification with the positron seems the most reasonable way of getting over the difficulty of the appearance of negative energies in our equations. On this view the positron is just a mirror-image of the electron, having exactly the same mass and opposite charge. This has already been roughly confirmed by experiment.*
>
> Paul Dirac in 'Theory of Electrons and Positrons', Nobel Lecture (1933)

As we have seen, the life story of the electron is also the story of the people involved in making discoveries associated with it – and few have been more eccentric than English physicist Paul Dirac. Born in Bristol in 1902 to Charles Dirac, a language teacher who had moved from Switzerland in his early twenties, and Florence Dirac (Holten), a librarian from Cornwall, Dirac's upbringing had unusual features. To encourage him to be bilingual, his father only spoke to him in French and his mother in English. Dirac claimed that, thanks to lack of contact with others while being home schooled, for several years he thought men and women spoke different languages.

An unusual life

The insistence of precision imposed on Dirac by his father from an early age may have influenced Dirac's scientific thinking. It has been suggested that Charles' requirement that Dirac only spoke perfect French to him or he would be punished resulted in a tendency to be taciturn, as Dirac would rather not speak at all than get things wrong. It can be misleading to attempt to diagnose medical conditions posthumously, but it does seem likely that Dirac was on the autism spectrum and this had an influence on his interpersonal skills – there was no doubt that he lacked any capacity for small talk and could be socially inept.

His upbringing was often an unhappy one. The family rarely ate together; despite being the younger of two sons, Dirac was picked out to eat alone with his father, while his brother and sister, considered intellectually inferior, were relegated to the kitchen with his mother. It seems to have been a sadly loveless time in his life. Dirac noted later, 'I did not know of anyone who liked anyone else.'

There is no doubt that in later life Dirac was liable to say unexpected things. The outgoing American physicist Richard Feynman noted that when he met Dirac, after the long, silent pauses that littered Dirac's attempts at conversation, his British colleague, attempting to be sociable, commented, 'I have an equation. Do you have one too?' Perhaps the best Dirac story, though, describes an incident when, on a lecture tour to the US, he had reached Madison, Wisconsin.

At one point in Dirac's talk, an audience member said, 'I don't understand the equation in the top-righthand corner of the blackboard.' There was an extremely long pause, after which Dirac's host, puzzled, asked him why he had not answered. Dirac responded, 'That was not a question, it was a comment.' Einstein would say of Dirac, 'This balancing of the dizzying path between genius and madness is awful,' while Dirac himself often introduced his wife Margit as 'Wigner's sister', rather than his spouse. This was because her brother was the Hungarian-American physicist Eugene Wigner, and for Dirac this was the most obvious way to put her into context.

Although Dirac began his academic life in Bristol, where he had gained degrees in engineering and mathematics by the time he was 21, it was when he moved to Cambridge to begin postgraduate work in physics that he flourished. He may have lacked social graces, but he opened up to the intellectual challenges and influences he would receive there. He had been supposed to work on relativity, but his supervisor, Ralph Fowler, was deeply engaged in the emerging field of quantum physics, and it was here that Dirac would shine.

While Dirac's biggest contribution would be to help bring together the originally conflicting ideas in quantum physics, derived from Heisenberg's matrix mechanics and Schrödinger's wave mechanics, as well as introducing relativistic effects to quantum particles, from the electron's viewpoint his chief importance was in imagining the existence of a new type of electron. If the electron is the hero of this book, Dirac was about to conjure up the anti-hero: the electron's nemesis.

From infinite sea to antimatter

Postulating the existence of a new particle would emerge from a bizarre consequence of Dirac bringing relativity into the world of the electron. Up to this point, quantum theory had avoided the oddities that emerge when particles move quickly, invoking Einstein's special theory of relativity.* But in 1927, Dirac formulated a four-part equation that took into account both the quantum nature of the electron and the impact of relativity.

Dirac's equation was considered remarkable ... but also problematic. As we have seen, quantum theory gave electrons a series of possible energy levels, which they could jump between by absorbing or giving off a photon of light. But there was nothing in Dirac's equation that allowed for this ladder of energy levels to have a lower limit: it appeared to show that it was perfectly possible for an electron to drop down to lower and lower energy, continuing even when that energy became negative.

---

* In the special theory of relativity, another product of his 1905 *annus mirabilis*, Einstein showed that when anything moves, from a static viewpoint, the moving object will increase in mass, shrink in the direction of movement and experience a slowing of time. These effects only become obvious when the object is moving very quickly. Quantum particles often do move very quickly, but up to this point, relativistic effects had not been factored into equations describing their behaviour.

Admittedly, this wasn't the first time that extremely effective physical equations had thrown up oddities. When the details of James Clerk Maxwell's equations for electromagnetic waves were explored, it was discovered that there were two solutions describing the progress of light. One described light as travelling forward in time from A to B, while the other had it simultaneously moving backwards in time travelling from B to A. This alternative solution was largely ignored as if it didn't exist (though we will see in chapter 30 that it would later be used to help understand an oddity of the electron). Similarly, one possible approach seemed to be to sweep Dirac's negative energy electrons under the carpet, simply ignoring them – but he came up with an alternative.

This took some time. By 1929, it had been shown that it wasn't so easy to ignore negative energy electrons using conventional theory. As a result, Dirac came up with a way that they could exist but be hidden away from sight. He envisaged a bottomless sea that was totally full of negative energy electrons. Because of this lack of space, although in theory an electron could drop down to a negative energy level, in practice it never would because all the places it could go to were full. As we have already seen when looking at atomic structure (see chapter 25), a fundamental law of quantum physics, the Pauli exclusion principle, discovered five years before Dirac came up with his negative energy sea, made it impossible for more than one fermion – the class of quantum particles including electrons – to be in the same state at the same time in the same system. If Dirac filled every possible negative energy state in his electron sea, there was nowhere for an electron to drop down below the zero-energy level.

While the existence of such a sea might seem a dramatically unlikely possibility, if Dirac's model was a good representation of reality, it should be testable, because there was a circumstance when a normal electron *would* be able to enter the negative energy sea – and if this happened, the electron would give off a detectable pair of photons. If a negative energy electron from the sea was stimulated to undergo a quantum jump to a higher level, then it would leave behind a gap – a 'hole' (as they would technically be called) into which a normal electron could drop. When Dirac thought about this concept, he realised that a missing negative energy electron with its negative charge would be identical to the presence of a positive energy particle that (unlike a normal electron) had a positive charge.

It was 1931 before Dirac was confident enough in his idea to write, 'A hole, if there were one, would be a new kind of particle, unknown to experimental physics, having the same mass and the opposite charge to an electron. We may call such a particle an anti-electron.' The only problem was that no one had ever seen such a particle, the first predicted example of what would become known as antimatter.

Missing the evidence

Towards the end of that year, Dirac headed off to Princeton University in America on a sabbatical. His timing could not have been worse, because at the same time, the familiar figure of American physicist Robert Millikan had turned up in Cambridge to give a lecture on cosmic rays. These are high-energy particles that crash into the Earth's atmosphere from space, which were a subject of considerable controversy.

The existence of cosmic rays had first been suggested by German physicist Theodor Wulf, who had taken a simple electrostatic detector (this was before Hans Geiger had developed his eponymous counter) up to the top of the Eiffel Tower in Paris in 1909. He claimed to have found a stronger background radiation level there than was present on the ground. At the time this was discounted, as it was believed that the only source of radiation detectable on Earth was radioactive materials in the ground.

Over the next few years, though, a range of experiments, many of them using balloons to take readings at high levels, suggested that there was more ionising radiation at altitude than on the surface.* This was even the case during an eclipse, ruling out the impact of the radiation from the Sun. This work would win the Nobel Prize in Physics for 1936 for Austrian-American physicist Victor Hess, who established the existing of this external radiation beyond doubt.

* Ionising radiation is radiation that ionises atoms when it strikes them. This can be both high-energy light (primarily X-rays and gamma rays) and energetic particles. Ionising radiation is the dangerous category of radiation, as it can cause damage to human cells.

The radiation was named by Millikan in the 1920s, who believed that cosmic rays were gamma rays from space, produced from nuclear fusion reactions, which produced electrons in the atmosphere by the process known as Compton scattering, where electrons are pushed out of atoms by high-energy photons. Unfortunately, Millikan was wrong here. It was soon discovered that particles from cosmic rays were far too energetic to be produced by Compton scattering – the actually cosmic 'rays' were showers of charged particles. We still aren't sure of the origin of all of these energetic particles, although some certainly seem to have originated in supernovas, massive stellar explosions. Some have originated outside the Milky Way galaxy.

Cosmic rays in the clouds

Part of Millikan's 1931 talk on cosmic rays in Cambridge included slides of cloud chamber images that were used to show particle paths. The cloud chamber has one of the oddest histories of any device in the physics laboratory, originating in a scientist's fascination with the mysteriously named Brocken spectre. This is not a German ghost, but a weather phenomenon associated with the Brocken, one of the mountains in the Harz range in Northern Germany. The Brocken has a connection with witchcraft in local legend, which may be why this eerie-looking apparition carries its name. In a Brocken spectre, a huge shadow of a person appears on mountain-top clouds, often featuring a glory, a rainbow ring around the shadow subject's head, like a multicoloured halo.

In 1894, Scottish physicist Charles Wilson had been on the UK's highest mountain, Ben Nevis, and saw a Brocken spectre, which stimulated an interest in the nature of clouds. This wasn't his day job, just a background activity, which explains why it took him seventeen years to perfect a device to help experiment with clouds. This would prove far more valuable when studying radiation than it was for meteorology.

The cloud chamber was a shallow, sealed cylindrical container with a clear glass top that was filled with warm, humid air. The volume of the chamber was then rapidly expanded using an attached diaphragm. As the pressure dropped, the water vapour inside formed a region of tiny water drops – in essence, a cloud. Wilson, who was working at

Brocken spectre.

the time on ionising radiation, noticed that when naturally occurring radiation passed through the chamber it left behind a trail of larger, visible droplets, seeded by the impact of the particle. He had discovered a way to see the invisible, thanks to the trail particles left behind.

Giving this work an early link to the electron's biography, Wilson was based at the Cavendish Laboratory in Cambridge, and one of his early tests, suggested by J.J. Thomson, was to use X-rays to produce a trail. Wilson would win the Nobel Prize for Physics in 1927 for 'his method of making the paths of electrically charged particles visible by condensation of vapour'.

Back with Millikan's 1931 Cambridge talk that Dirac so unfortunately missed, some of his slides were of cloud chamber images taken by his graduate student Carl Anderson as cosmic rays passed through a cloud chamber. To help understand the nature of these 'rays', the experiment placed the chamber in a strong magnetic field. Particles that were electrically charged would display a curved path – the direction of curvature showed if they were positively or negatively charged, while the degree of curvature depended on the momentum and amount of charge.

As cosmic rays travel through the chamber and collide with the molecules inside it, they can produce a spray of other particles, so-called secondary cosmic rays. Anderson noticed that when electrons were produced (identifiable from their distinctive path shape), they were sometimes paired with another particle. At the time the images weren't good enough to be sure, but by the following year Anderson had a single, good enough image to clearly show that a collision had resulted in the production of an electron along with another particle following a mirror image path, suggesting that it had the same mass and charge magnitude as the electron, but was positively rather than negatively charged.

Anderson wondered if someone had been playing around with his powerful electromagnet, reversing its polarity partway through the exposure, but over the next few weeks he managed to find two more

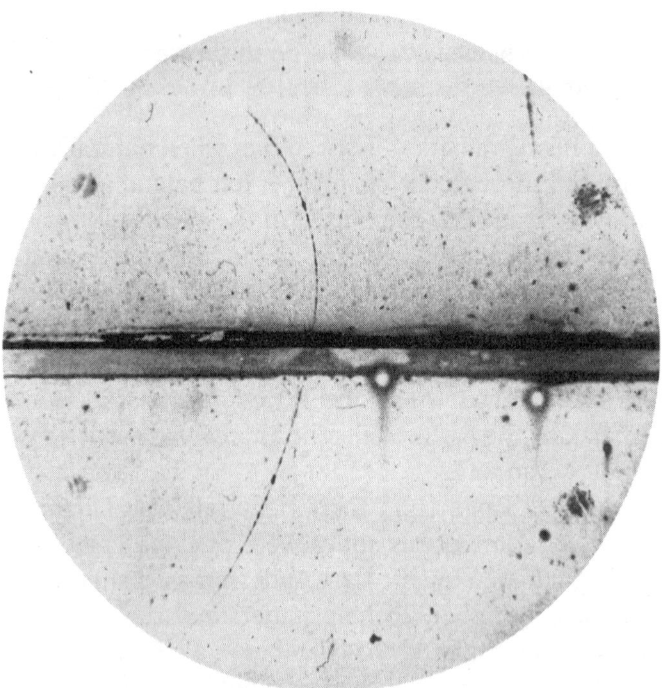

Anderson's positron image.

Carl D. Anderson (1905–1991)

Public domain, via Wikimedia Commons

of these positive electron-like tracks. He did not, however, associate the evidence with Dirac's theory. A first response was to assume experimental error, but the American mathematician Rudolf Langer suggested that Anderson had spotted Dirac's anti-electron. No one took any notice.

By the end of 1932, a new, state-of-the-art cloud chamber had been set up at the Cavendish Laboratory in Cambridge. The big advance made by Dirac's friend Patrick Blackett and his colleague Giuseppe Occhialini was to make chamber photography more efficient. With earlier devices almost all the photographs were blank, as they simply took a shot and hoped a cosmic ray had made an appearance. Blackett and Occhialini used Geiger counters to spot incoming cosmic rays, which were used to trigger the camera, making the process far more effective. They soon found their own positive electron. Up to now, Dirac hadn't heard of the discoveries, but he sat through a presentation by his colleagues, seeming not to make a direct connection.

It wasn't until the next year that the collected evidence had grown to such an extent that it became clear what was being seen was an anti-electron. Anderson would be awarded the 1936 Nobel Prize for Physics for discovering the positron (alongside Victor Hess for his cosmic ray discovery). This name didn't come from Anderson, but was suggested by the editor of *Physical Review*, the journal where Anderson published his discovery paper.

Positrons in action

What was happening in the production of electron–positron pairs was a demonstration of the reality of $E = mc^2$, the iconic formula Einstein derived from his special theory of relativity, showing the relationship between the mass of matter and energy. The incoming cosmic ray released enough energy in its collision with atmospheric molecules to produce twice the mass of an electron, in the form of this pair of quantum particles.

Other experiments demonstrated an outcome that was predictable from the onset using Dirac's negative energy sea model. If a positron was the equivalent of a hole in the negative energy electron sea, then an ordinary positive energy electron could drop down into that hole.

As it dropped in energy and disappeared into the negative energy sea it would give off a pair of photons. The mass energy of the electron and the positron particles would blast out of this interaction as this high-energy electromagnetic radiation. The original pair of particles would disappear – annihilated in the jargon.

The concept of antimatter was soon expanded to include antimatter equivalents of the other known particles. Antimatter is now routinely made in (extremely) small quantities at facilities like the European CERN laboratory. Matter–antimatter annihilation is potentially dangerous – but from the viewpoint of the life story of the electron it was like the discovery of a long-lost twin. Just twins that should never be allowed to meet.

The positron has now escaped from the laboratory to take part in one of the scanner types available to modern hospitals. In a PET scan (Positron Emission Tomography), the subject is injected with a radioisotope – a material in this case which undergoes a variant of beta decay called beta+ or positron decay. Here, instead of a neutron decaying to produce a proton, an electron and an antineutrino, a proton is converted into a neutron, releasing a positron and a normal matter neutrino.

The positrons released in the body annihilate with electrons to produce pairs of gamma ray photons originating inside the body, which are detected by a ring-shaped scanner. Although gamma rays are dangerous ionising radiation, the level produced is sufficiently low to be of minimal risk.

By the time the positron had been discovered, quantum physics was moving on. After a considerable amount of confusion in the early years, a picture was emerging that would provide a much better understanding of the interaction of matter, particularly electrons, and particles of light. This theory would be known as quantum electrodynamics, or QED for short.

QED

29

I'm going to describe to you how Nature is – and if you don't like it, that's going to get in the way of your understanding it … The theory of quantum electrodynamics describes Nature as absurd from the point of view of common sense. And it agrees fully with experiment. So I hope you can accept Nature as she is – absurd.

Richard Feynman in *QED: The Strange Theory of Light and Matter* (1985)

QED: not *quod erat demonstrandum*, but 'quantum electrodynamics'. As quantum theory matured, one of the most important aspects to be explained and understood was the interaction of light and electrons. The Nobel Prize in Physics for 1965 was awarded to three physicists for their development of this field: Sin-Itiro Tomonaga, Julian Schwinger and Richard Feynman, with the rather poetic description of being for 'their fundamental work in quantum electrodynamics, with deep-ploughing consequences for the physics of elementary particles'.

Three basic actions

QED and its offspring QCD (quantum chromodynamics, which deals with the interactions of the quarks and gluons that make up protons and neutrons) are not just about the electron, so we only need concern ourselves with one aspect of this where electrons are involved. American physicist Richard Feynman says in his book *QED** that there are just three basic actions from which all the phenomena involving light and electrons arise:

- A photon goes from place to place
- An electron goes from place to place
- An electron emits or absorbs a photon

Feynman, born in New York in 1918, remains one of the most revered physicists in the trade. Spurred on by his father Melvyn's interest in the world around him, Feynman was taught from an early age that understanding nature is about more than simply giving things labels. While at graduate school, he attended undergraduate biology lectures. When given the task of explaining electrical impulses in the nerves of a cat, Feynman began by explaining the muscular structure of a cat. Subsequently told by the undergraduates that they already knew what all the muscles were called, Feynman claimed he responded, 'Oh, you do? Then no wonder I can catch up with you so fast after you've had four years of biology.'†

Feynman's physics career would be sidelined for a while by the Second World War, where he became a major contributor to the Manhattan Project, working on the development of the atomic bomb. When not involved in devising methods to enrich uranium, or to work on calculations to predict the behaviour of nuclear materials reaching criticality, his lifelong distaste for box-ticking conformity led him to repeatedly test the security applied to top

---

* See Further Reading.

† I say Feynman claimed to have made the comment about catching up so fast with the biology because, by implication, they had wasted time learning unnecessary labels, as he had something of a reputation for embroidering his stories – but whether or not it exactly happened this way, it beautifully illustrates his approach.

secret documents, picking locks on secure filing cabinets as a mental challenge.

Although he never wrote a book, books based on Feynman's lectures and anecdotes have become bestsellers. With his broad New York accent and anything but laid-back lifestyle, Feynman was very different from the idea of an academic at the time. He had a flair for the dramatic, most notably when on the 1986 *Challenger* space shuttle commission, uncovering the cause of the disaster. Escaping the controlled visits to suppliers, Feynman discovered that the rubber O-rings used to seal joints in the rocket motors lost flexibility at low temperatures. Unwilling to wait for the bureaucratic delays, he used a televised session of the commission to make this fault visible by plunging a section of O-ring into his iced water and dramatically demonstrating its lack of flexibility.

The quantum diagram

Although Feynman, Schwinger and Tomonaga all came up with equivalent but different formulations of QED, what made Feynman's contribution particularly valuable was the development of Feynman diagrams. These make use of simple graphical symbols – for instance, straight lines for the paths of moving matter particles and wavy lines for photons – to show the interactions between quantum particles. The diagrams, as we shall see, are far more than simply a visual representation of the interaction between particles – they can be used as the basis for calculations, where the size and direction of each line corresponds to a mathematical formula, uncovering the complexities of quantum probabilities.

When thinking of the interactions of electrons, it might seem that QED would be limited to explaining quantum technology such as photoelectric cells. But the photon is not just the particle of light. In quantum physics, forces are communicated by carrier particles. These particles are often not directly observed (hence they are often referred to as virtual particles), but make it possible for one matter particle to influence another.

So, for example, when one electron repels another electron, because they have the same electric charge, they don't come in direct contact with each other. The repulsive force acts on the two electrons at a distance because they are sharing photons, which are the carrier particles for electromagnetism. And Feynman's diagrams make it easier to get a feel for how this QED interaction occurs.

A simple Feynman diagram like the one below shows such a repulsion occurring:

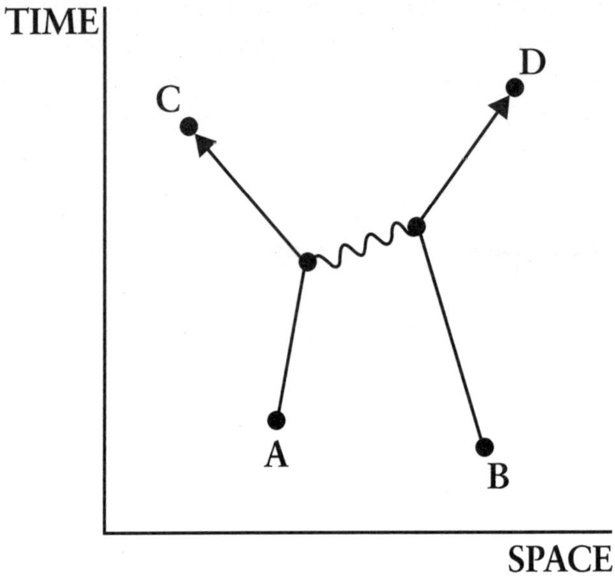

Electron repulsion in a Feynman diagram.

Two electrons start off at the points labelled A and B. As time elapses (moving up the diagram), they initially head towards each other before exchanging a photon that results in them repelling each other and heading off towards C and D respectively.

Because a quantum interaction outcome is based on probabilities rather than certainties, there are many more ways that a Feynman diagram can be used to describe even a simple electron interaction like this. For example, all the diagrams below are among those representing the interaction of two electrons.

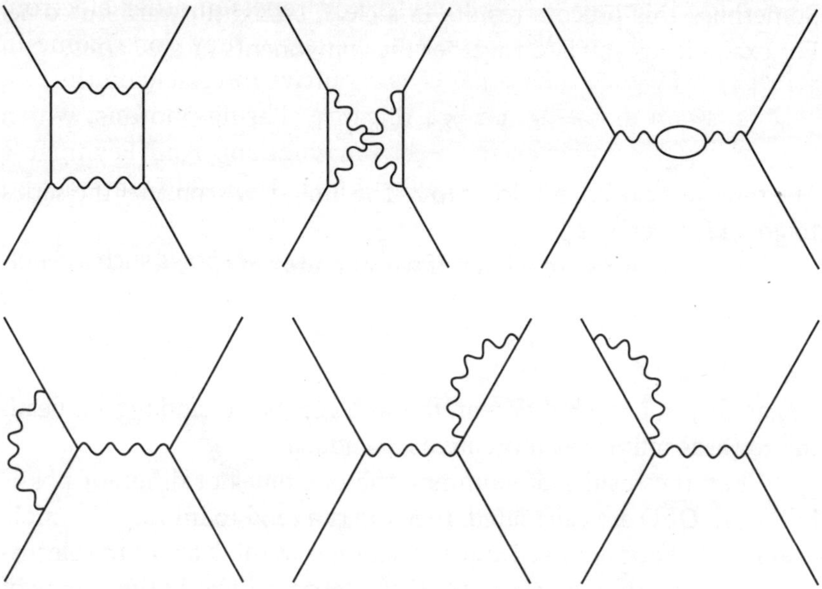

More complex electron repulsion in Feynman diagrams.

This approach can be carried on with more and more complexity, but as extra levels of interactions are included, the probability drops. With a relatively small number of diagrams in a simple interaction like this it is possible to achieve high levels of accuracy. Before long, quantum theory makes incredibly accurate predictions of the outcomes. Feynman pointed out that the theory's predictions were so close to what was observed in experiments that it was the equivalent of predicting the distance from New York to Los Angeles with the difference between reality and prediction being the width of a human hair.

To infinity and beyond

Although QED did have the accuracy that Feynman claimed, it also came with a built-in problem. In effect, combining possibilities from different Feynman diagrams with attached possibilities is a bit like the common mathematical process of summing an infinite series.

Sometimes this process results in a clear, straightforward outcome. For example, as you add together the components of the series:

$$1 + \frac{1}{2} + \frac{1}{4} + \frac{1}{8} \ldots$$

The total gets closer and closer to 2. The limit, if we consider the series to go on forever, is 2.

But some series are wilder. If we sum up:

$$1 + \frac{1}{2} + \frac{1}{3} + \frac{1}{4} + \frac{1}{5} \ldots$$

Apparently not much different, the sum gets bigger and bigger, heading towards infinity as more items are added.

When the results of summing the outcomes of different possibilities in QED are calculated, they too can tend to infinity. A simple example that doesn't need any calculation is to think about the electrical field strength of an electron. This is proportional to the square of the distance to the centre of the electron. If you half the distance, you quadruple the strength. But in theory an electron is a point particle with no diameter at all. You can get as close as you like, leading to an infinite value.*

Given that QED was so effective, with the exception of these tendencies to run away with itself, a key part of making it acceptable was to overcome any theoretical infinities that cropped up in the calculations. In the real world, these infinite values were not observed. As a result, in a process that became known as renormalisation (but might better have been called fudging), observed values were plugged in to replace infinities where required, leaving QED as an impressive predictor of reality beyond those observed values.

Over time, mathematical methods were introduced to make it less necessary to rely on renormalisation, working around the possibilities

* In QED proper it is a little more complex when calculating the electric field strength at a particular location with respect to an electron as the electron itself doesn't have a location, just probabilities of being in different locations, which requires a calculation combining probabilities and locations – but the result still has a worrying potential for infinite outcomes.

for infinities to occur. It is generally acknowledged that when an infinite value emerges in a calculation there is something not right with the calculation's match to the real world. Sometimes this can lead to compromises. For instance, a black hole conceptually involves an infinite value at its core.

Although astrophysicists talk as if black holes exist just as theory predicts, what they really mean is that there are observed bodies in space which behave as theory tells us black holes should. But because we can never observe what is happening inside them, we don't know exactly what is occurring. All we can say for certain is that theory predicts an infinite gravitational field, but must be wrong about this. In a similar way, QED makes very accurate predictions, but the reality is subtly different in a way that may never be fully understood.

In parallel with the development of QED, some electron-related oddities would be discovered, both in the behaviour of positrons and in the way that a problem with electron recoil would suggest that a common physical process involves photons travelling backwards in time.

TIME TRAVEL, ELECTRON STYLE

30

The radiation by black holes shows that quantum theory allows travel back in time on a microscopic scale and that such time travel can produce observable effects.

One can therefore ask: does quantum theory allow time travel on a macroscopic scale that people could use?

Stephen Hawking in *A Brief History of Time* (1998)

Although most of our ideas about time travel are nothing more than (entertaining) science fiction, there are elements of the behaviour of electrons where for theoretical reasons we appear to see backward time travel. This is the interesting side of time travel. Travelling into the future is easy. Just wait a few seconds and there you are. You can also speed this process up. If you travel extremely quickly with respect to the Earth, for instance, when returning home you will find you have travelled into the planet's future. It takes speeds we aren't currently capable of to make this noticeable, but Einstein's special theory of relativity makes it inescapable.

Electrons go back to the future

In itself, though, forward time travel is nowhere near as interesting as heading for the past – it's here you get to send lottery results back to yourself, or to set up bizarre paradoxes.* And there are (at least) two distinct aspects of the nature of electrons where this arises. The first can be seen in the process known as scattering. This is a very common phenomenon where a photon is absorbed by an electron, boosting its energy. Some while later, the electron re-emits a photon, which could have a different energy (colour) or head in a different direction.

One kind of scattering, Rayleigh scattering, is the reason that the sky is blue. Photons from the sun are absorbed by electrons in atoms in the atmosphere, then re-emitted. Because higher energy photons are more likely to be absorbed, the blue end of the spectrum does this more and heads from the sky down towards us. Another form is Compton scattering, where a high-energy photon pops an electron out of an atom. In a Feynman diagram, a typical example of scattering might look a bit like this:

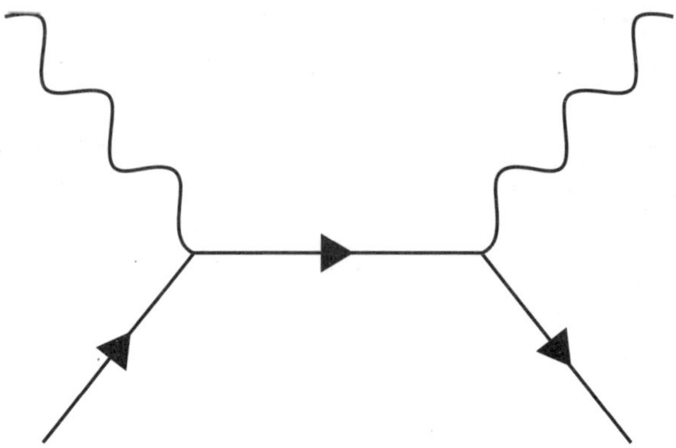

Scattering in a Feynman diagram.

That is one possibility for scattering, but bear in mind that in quantum theory, with differing probabilities, interactions between particles can happen in many different ways. Here is another perfectly feasible Feynman diagram for scattering:

---

* The most commonly cited backward time travel paradox is the so-called grandfather paradox, where you go back in time and kill a parent or grandparent before you are born. So now you don't exist. Which means it never happened (and so on). I much prefer the self-writing book. Imagine I could send the manuscript of this book back to myself at a point in time before I started writing it. In the past, I then send it off to the publisher. Where did the book come from?

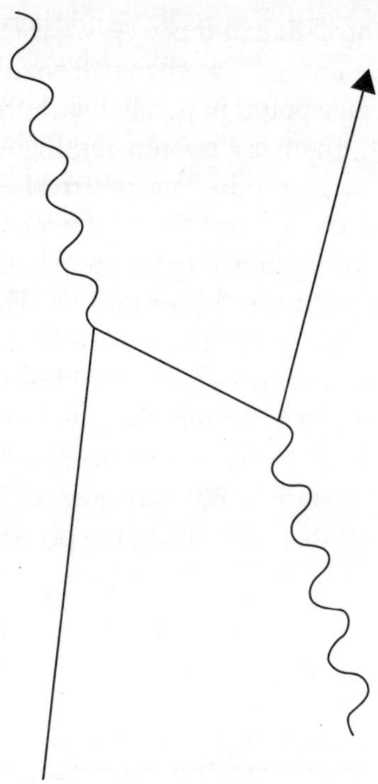

The time travelling electron.

Remember that in these diagrams, time is increasing up the page and movements in space are on the horizontal axis. If we start at the bottom left, we see an electron moving through space and time. After a while it emits a photon. As a result it changes direction and travels a little further before absorbing another photon, causing its final change of direction. But take a closer look at that middle section. As well as going left to right, it is going downwards. That is backwards on the time axis. It's a time traveller.

It's entirely possible for something to be compatible in theory but never to happen in practice. Yet this is a common enough interaction. Why, then, do we not have excited newspaper headlines about time machines? Because from our familiar viewpoint where time moves forward, what we see is a type of electron that would be attracted rather than repelled by another electron. In QED terms, an electron travelling backward in time is the same thing as a positron travelling forward in time. If we see that central line in this way, what is happening totally shifts, like one of those images that change from a duck to a rabbit.

With the positron viewpoint, in parallel with the electron coming in from the bottom left, there is a photon coming in from the bottom right. The photon's energy produces a positron/electron pair, with the positron heading to the left and the electron heading to the right. Both are moving forward in time. Finally, the positron meets up with the electron coming up from the bottom left. The two annihilate, producing the photon that heads off towards the top left.

In a sense, this example of time travel could be considered a cheat, though it would be a mistake to say that the 'positron travelling forward in time' interpretation is any more correct than the 'electron travelling backward in time' interpretation. Both are entirely legitimate, it's just that we are more comfortable with taking the positron viewpoint.

Unleashing advanced waves

The other example of time travelling caused by electrons, though, is not just a matter of interpretation. It is a mechanism by which, in theory, it would be possible to send messages back in time. The only catch with this one is the 'in theory' part. Because there is a requirement to make this work that may well be practically impossible to fulfil. Once again, we have the involvement of Richard Feynman, this time with John Wheeler, Feynman's supervisor when a young scientist.

Wheeler, born in Jacksonville, Florida, in 1911, is not as well-known as Feynman, but he was a major influence on physics in the second half of the twentieth century. Like Feynman he was involved in the Manhattan Project before working on the fusion bomb, then reviving work on Einstein's general theory of relativity, which had become less popular as quantum theory took the attention of the leading physicists. He is often said to have come up with the term 'black hole', though in reality he only popularised it.*

* 'Black hole' is thought to have been first used at a 1963 astrophysics symposium, though no one is sure who coined it. It first appeared in print in a 1964 report on the event by journalist Ann Ewing. John Wheeler started to use it in 1967 and made it the standard term.

Wheeler and Feynman were thinking about an oddity that emerges from Maxwell's equations that describe the working of electromagnetism. Sometimes an equation can have more than one solution. The chances are you had to solve a quadratic equation at school, something like $2x^2 + 10x + 12 = 0$. With a bit of mental wrangling, you can see that this is the same as $(x+3)(2x+4) = 0$ – because when you multiply the two bits in brackets together you get the original equation.

The second form of the equation makes it clear that if x is either -3 or -2, then one of the bracketed parts is zero, so the correct result is obtained. This means that there are two solutions for the equation, each equally valid. Similarly, Maxwell's equations for the production of light can be solved in two ways. These are given the names of 'retarded' and 'advanced' waves. Retarded waves are what we usually see. Imagine we send a beam of light from point A to point B. The light sets off at midday and arrives at its destination one second later. This is a retarded wave. But the equations also say there should be an advanced wave that sets off from B at one second after midday and arrives at A one second earlier.

There is no physical reason why advanced waves should not exist – they just can't usually be observed. Yet if it were possible to detect them, we could use an advanced wave to send a message from Earth to a distant location in space – say a satellite around Jupiter. On average, it takes light around 40 minutes to cover this distance, so the message would arrive at Jupiter 40 minutes before it was sent. Use another advanced wave to send the signal back to Earth and it would get there 80 minutes before the original message was broadcast.

When advanced waves were first noticed as a feasible solution to the equations, it was assumed that they could just be ignored, keeping only the retarded wave solution. But as there is no physical reason to do this, some have always thought that advanced waves could exist undetected. Undetectable reality is common in quantum physics. For example, as we have seen, photons carry the electromagnetic force between particles. There is a constant interchange of photons inside an atom between protons in the nucleus and electrons, keeping the electrons from escaping. These are called virtual particles, because they are never observed in normal circumstances. But there are circumstances when we can expose a photon, disrupting its normal invisibility. Could it be the case that there would be a possibility of exposing advanced waves? Wheeler and Feynman proposed to do this for the interaction between an electron and a photon that it emits while in an atom.

Dealing with electron recoil

Although photons have no mass, they do have momentum.* Like energy (and a number of other things) momentum is conserved. If you give something forward momentum, it will give you equal and opposite backward momentum: this is implied by Newton's third law of motion. Perhaps the most familiar example of this is recoil. When, for example, a gun gives forward momentum to a bullet, the gun (and hence the person firing it) feels a backward push. Similarly, when a photon leaves an atom, the electron (and hence the atom) feels a recoil in the opposite direction.

However, unlike a gun and a bullet, the electron is electrically charged and interacting electromagnetically with the nucleus of the atom. The recoil causes the electromagnetic field of the atom to act on itself, producing just the kind of spiralling to infinity of calculated outcome that plagued QED. Wheeler and Feynman found a simple way to deal with this problem – because it was only a problem if you ignored advanced waves. As a starting point we have to think about what happened to the photon after the electron emitted it.

Let's imagine that photon shooting away from an atom. It might encounter an air molecule or a solid object and be absorbed by an electron. It might reach empty space and travel for thousands of years. But at some point it is likely to hit something. According to Wheeler and Feynman's 'absorber theory', we can now bring advanced waves into play. Although we don't see both, they suggested, there are two

---

* If a vaguely remembered value for momentum from school of mv (mass times velocity) troubles you in imagining that a massless photon can have momentum, there is an additional component to momentum which arises from the relationship between mass and energy that comes from special relativity. Photons do have energy, and the momentum can be calculated from that energy divided by the speed of light.

photons involved. The one we have already mentioned travelling forward in time, and the advanced wave photon, which leaves the eventual point of impact (the absorber in the theory's name) at the time the standard photon arrives, travels backward in time and gets to the point of origin as the standard photon leaves.

When that advanced wave photon arrives at the source, it hits it, causing the apparent recoil. There is no longer the self-interaction generated by assuming that the electron is recoiling from the departing photon. It was a separate photon that started the movement. Each of the two photons would have half the total energy of the interaction, but they would both always be in exactly the same place, one moving forward in time, the other backwards. As a result, the total energy at their location would always be correct.

Absorber theory is rarely given much consideration, not because it's wrong, but because it is effectively indistinguishable from the theory that ignores it, apart from doing away with self-interaction. But arguably it is more elegant, bringing as it does both solutions of Maxwell's equations into play. It is just something that some physicists find philosophically troublesome because of that time-travelling photon. To others, however, it most accurately describes the relationship between an electron in an atom and a photon.

A direction without absorbers

What gets really interesting, though, is if we consider that requirement for there to be an absorber. If the theory is correct, then things would be different if the photon never hit anything, but simply carried on forever. It is only when the photon is absorbed that the advanced wave photon can head back in time.

Imagine, then, that we found a direction in which there were no absorbers. Ever. First, we need to send a good, absorbing object out a good distance into space in that direction. We then arrange for that absorber to be moved in and out of a stream of photons sent from Earth. When the absorber is in the stream, everything will work as normal. But pull the absorber out of the way and the photons could no longer leave Earth as they are not going to be absorbed.

However, because the reaction from the absorber moving registered on Earth is due to the appearance and disappearance of photons travelling backwards in time, the movements of the absorber provide a signal that travels back in time.

In practice, to make this a viable way to communicate with our past, it would be necessary to have an equivalent route from the absorber towards the Earth that misses it and never gets absorbed. Then the reverse process could be made by introducing an absorber from near the Earth so a signal would go back in time to the distant location, then back again when returning. This would add a little positive time to the transmission as the signal would have to be relayed between Earth and the absorber station using a conventional means – but it is unlikely to wipe out the whole of the effect.

As is usually the case with theoretical time machines, there is a limit on how far back you can send a message. Even in theory, it's not possible to send something back to before the time machine was first created (so even if there were a time machine that transported more than information, it would not be possible to visit historical events that came before that date). In this case, the transmitter has to be working before we can see how it responds to moving the absorber.

Time travelling as a result of electron interactions is a fun topic to play with, but as yet has not proved a practical use of our understanding of the quantum nature of electrons. But this doesn't mean that electrons themselves cannot be harnessed for new uses. We saw in chapter 24 how electronics began with the use of thermionic valves. But this development of electron abilities would be transformed and made central to everyday life by moving from vacuum tubes to a solid state alternative.

31 GOING SOLID

I would like to start by emphasizing the importance of surfaces. It is at a surface where many of our most interesting and useful phenomena occur. We live for example on the surface of a planet. It is at a surface where the catalysis of chemical reactions occur. It is essentially at a surface of a plant that sunlight is converted to a sugar. In electronics, most if not all active circuit elements involve non-equilibrium phenomena occurring at surfaces.

Walter Brattain in 'Surface properties of semiconductors' – Nobel Lecture (1956)

The early electronics first gave the electron a starring role, requiring engineers to have a conscious awareness of its existence, rather than simply dealing with it as a background actor. But we can date the particle's true moment of breaking out to a wider world to 1947 with the first demonstration of a working transistor. If we think about the electron taking part in modern everyday lives beyond its hidden roles in atoms and electricity, it is in solid state electronics, be it in our phones, the computer I'm typing this on, our cars, or the internet servers behind an AI chatbot – the whole infrastructure of our modern electronic lives.

The thermionic valves that were central to the first electronic computers had four main issues. They were fragile – easily broken when dropped or shipped. They regularly burned out as the heated electrode lost material. They were large – typically the size of a saltshaker and sometimes larger than a fist. And they used a lot of electricity. When

SF author James Blish wrote his 1956 novel *They Shall Have Stars*, he looked forward to a time when spaceflight to Jupiter would be readily achieved. But when humans arrived there, he noted that 'there was no electronic device ... since it was impossible to maintain a vacuum on Jupiter'. The atmospheric pressure of the giant gas planet, he assumed, would crush the valves with their evacuated interiors. For electronics to become as central to our lives as they now are, there needed to be components that were robust, long-lasting, small and low consumers of energy.

Semiconductors at Bell

The breakthrough required happened at what could be argued was the second of the electron's two spiritual homes, after the Cavendish Laboratory at Cambridge. This second location was Bell Labs, part of the American communication giant AT&T based at Murray Hill, New Jersey, where it split off from the company's West Street base in 1941. Bell Labs (now part of Nokia) was AT&T's incredibly innovative research arm. The starting point towards the transistor was an understanding of semiconductors and of the quantum behaviour of electrons.

Most of us are familiar with the term 'semiconductor' in the context of electronics, even if exactly what the term means feels a touch uncertain. Conductors (as far as the electron is concerned) are materials through which electrons are able to travel, while insulators stop them dead. It's not exactly obvious how a substance can be a 'half' conductor, with the implication that it conducts ... but doesn't. Yet that's exactly what the name suggests.

The best-known semiconductor elements are silicon, germanium and selenium. Their special power when it comes to electrons is that they will conduct in some circumstances and not in others – exactly what's required if we are to reproduce the abilities of a thermionic valve (though semiconductors are more versatile in the ways that these circumstances apply).

The distinction between conductors, semiconductors and insulators lies in what's known as the conduction band for electrons. As we have seen, the electrons that inhabit an atom live in clouds of probability known as orbitals. In some solids, outer orbitals of adjacent atoms can be close enough to each other to run together, providing a kind of continuous electron highway through which the particles can flow – the conduction band. But in an insulator, there is a big 'band gap' between the atoms' orbitals and this highway, meaning that no conduction can occur. A semiconductor also has a band gap, but it is narrow. If the outer electrons can be given a bit of a push, they will enter the conduction band and become able to flow.

In some materials, the jump across the narrow band gap might happen when energy is gained from incoming light, making the semiconductor light sensitive. In others, 'doping' by adding a small amount of another element can result in the addition of extra electrons or 'holes' to encourage flow.

Holes here are effectively non-electrons. If you imagine a stream of electrons with gaps in it flowing through a conduction band, these gaps are referred to as holes. In a semiconductor where electrons are flowing in one direction through the conduction band there tends to be a corresponding flow of holes in the opposite direction in the band below it, known as the valence band.* Semiconductors can be doped with electron-rich materials (n-type doping) or hole-rich (p-type).

Three figures from Bell Labs would share the 1956 Nobel Prize in Physics for 'their researches on semiconductors and their discovery of the transistor effect' – they were William Shockley, John Bardeen and Walter Brattain. Shockley was the manager who set the project of developing a solid-state equivalent of a triode valve underway. Bardeen was a theoretical physicist, who understood the quantum implications of using these materials, while Brattain was the engineer who built the prototype.

* Although the terms 'valence' and 'valence band' are related, they are not the same thing. Valence describes how an atom will combine with other atoms to form molecules, usually based on the number of bonds needed to fill the atom's outer electron shell. Both the valence band and the conduction band will usually be linked to the outermost shell.

The solid-state triode

The first transistor looked more like a prop from a low-budget science-fiction film than a compact electronic component. Including its supporting structure, it was about the size of a hand, consisting of a flat metal electrode with a grey germanium crystal on top of it. The germanium was in two layers, the top of which was p-type and the bottom n-type. Above the semiconductor crystal was a triangle of plastic with gold foil on the outside. The foil was broken with a narrow gap at its point, which was pressed against the germanium by a spring. The result was that a small current between one of the gold foil pieces and the electrode at the base controlled a much larger current between the other piece of foil and the electrode.

The name for this clumsy looking device (which was only a demonstrator – never intended for practical use) was the result of an internal competition at Bell Labs in the days when this kind of exercise could come up with something better than Boaty McBoatface. There was little hope of a win for entries like 'surface state triode' or 'semiconductor triode', but John Pierce's 'transistor' had hot competition from another name that echoed many pieces of scientific hardware of the period: 'iotatron'. It's probable that transistor won because it more closely paralleled the name of existing electronic components: resistors, capacitors, varistors and thermistors.

Shortly after the successful demonstration in December 1947, William Shockley attempted to steal his team's thunder. He had been working on a different kind of solid-state triode for some time, but had been unable to get it to work. Shockley horrified Bardeen and Brattain by announcing that he, rather than they, should be named on the patent for the transistor, as he had provided the groundwork with his failed device. As Shockley put it (according to Brattain), 'Sometimes the people who do the work don't get the credit for it.' Bardeen seemed stunned into silence by Shockley's announcement, but Brattain shouted, 'Oh hell, Shockley, there's glory in this for everyone!'

Perhaps reflecting the hierarchical nature of business and research facilities at the time, Bell Labs lawyers seemed to fall in behind Shockley's claim, until it was discovered that an existing patent for another failed device was too similar to Shockley's idea to make it feasible to register a patent using this basis. As a result, Bell then switched to specifically applying for a patent that named Bardeen and Brattain, more reasonably based on their successful approach.

If this had been the end of the story, Shockley would probably have come out of things badly. (As it was, his reputation never entirely recovered.) But while a development of Bardeen and Brattain's point-contact transistor did go into production, it was not easy to miniaturise and was physically clumsy and fragile. Shockley, inspired by the pair getting something working, came up with an entirely new design, the junction transistor, less than year later. This did away with the fiddly point contact, instead using a three-way sandwich of P-N-P doped semiconductors, variants of which would dominate electronics for two decades.

The amazing shrinking component

At this stage, while smaller, cooler (in both senses of the word) and less fragile than a thermionic valve, transistors were still relatively large, the size of a mid-sized coin, and depended on the expensive element germanium to provide semiconductor magic. The next major step on the way to modern electronics was the move in 1954 to begin using the far cheaper and ludicrously abundant silicon.*

One further step was required to move from transistors being chunky individual components to the mass-inhabitants of printed circuits on chips we know today. At the time of writing, the processor alone in the latest smartphone contains between 15 and 20 billion transistors – doing this with individual transistors would require a monster device that would have made ENIAC (see page 189) look tiny.

* Silicon dioxide, also known as silica, makes up the mineral quartz and is the main component of most types of sand. Silicon itself is the eighth most abundant element in the universe, and the second-most abundant in the Earth's crust after oxygen. Rare, it is not – though producing suitably pure silicon wafers is not trivial.

Shockley's failed solid-state component had been an attempt to create a triode using the field effect. The idea was that the electrical field from an electrode referred to as a gate would be used to influence the flow of current in an adjacent semiconductor. The problem that Shockley and others faced in making this work was that quantum effects in a semiconductor usually prevent an external electrical field from being able to influence it.

This electronic roadblock was overcome by accident in 1955, when a thin layer of silicon dioxide deposited on a wafer of silicon crystal was discovered to make it possible for an electrical field to penetrate the semiconductor. By 1959, the metal-oxide semiconductor field effect transistor (MOSFET) that became the standard had been developed using this approach. The ability to use a thin oxide, later on a silicon wafer, opened up the possibility of developing integrated circuit chips, the start down the road of miniaturisation that makes the wonder that is a modern phone possible.

Electrons get flash

One other quantum trick up the electron's sleeve would later transform another essential for effective computing – storage. From the earliest days of computers, it was apparent that information would need to be held somewhere for future use. Storage began mechanically, with holes punched in paper strips and cards, before going magnetic, with first tapes and then the much more flexible disk drives. Although the technology advanced hugely, along with external storage on optical disks, these would remain standard for computers until relatively recently.

But just as you could not have a smartphone with a host of individual transistors inside, so too it wasn't practical for a phone to contain a disk drive, which is relatively large and easy to damage if moved quickly. Instead, phones rely on flash memory, where electrons make use of one of the more surprising implications of quantum physics, known as tunnelling.

Flash memory was invented at the Japanese firm Toshiba around the beginning of the 1980s and was originally primarily hidden away and only indirectly accessible; it was used to hold information that was rarely changed, like the BIOS* instructions that a computer uses when it first starts up. This was because early flash memory was expensive and slow to read and write, so not ideal for everyday rapid memory use.

The first of the new generation of flash memory chips came into use during the 1990s, in removable memory cards, and these days this memory which holds its data without power provides the storage in phones and compact computers, where we take it for granted that many gigabytes of information can be safely stored on a small, portable device. This type of flash memory is much quicker to access than the earlier version. It does have a limitation, though – it can't access a single location at a time, instead pulling in or writing hundreds or thousands of bits simultaneously – but this is easily worked around, is sometimes useful and is always far outweighed by the benefit of speed of access.

Flash memory, like the conventional type of storage, makes use of transistors, but these are special variants of MOSFET devices called floating-gate transistors. As we have seen, the gate is the transistor's equivalent of the grid in a triode, the electrode that controls the flow through the thermionic valve. In a floating-gate transistor, there are two gates: a traditional 'control' gate and, beneath it, a floating gate which is electrically isolated so it can hold a charge indefinitely, providing the data storage. When the floating gate is charged up, it screens the control gates away from influencing the flow through the transistor, giving it a permanent switching role.

The floating gate is totally isolated by insulators – which leaves the problem of how to change the value of this inaccessible memory by charging up or erasing the charge on the gate. And it is here that a purely quantum effect takes over. Charge is added to the gate, or wiped from it, using quantum tunnelling: this is a process by which a

---

* BIOS stands for Basic Input/Output System – it is, in effect, a tiny operating system to get things on the computer going enough to be able to kick off the real operating system and to provide very low level services, for example responding to the press of a key.

particle – in this case, an electron – can pass from one side of a barrier to the other without passing through the space in between.

Tunnelling happens because of the fuzzy nature of the location of a quantum particle until it interacts with its environment. Schrödinger's equation tells us that we can only give probabilities for finding a particle in various locations, probabilities that spread spatially over time. This means that a quantum particle can already be on the other side of a barrier it can't get past, enabling it to interact. Without explicitly exploiting this weird quantum effect, this type of floating-gate transistor could not function. Within any device using flash memory for storage there is a whole bunch of tunnelling going on.

As flash memory became more effective and cheaper, it has gradually sidelined other means of storage. We now transfer data using memory sticks (thumb drives) rather than CD-ROMs. And most new desktop and laptop computers are likely to have SSDs – solid state disks which are large-scale flash memory stores – rather than magnetic disk drives.

When first exploring semiconductors, we noted that in some cases light could be used to push electrons up into the conduction band. This possibility led to another of the electron's crucial twentieth century and beyond technologies. Devices that could convert sunlight into electrical energy.

ELECTRONS FROM THE SUN 32

In all previous cells, so far as I am aware, the two portions or parts of the selenium, at which the current enters and leaves it, have been in substantially the same electrical state or condition. Furthermore, the paths of the current and of the light have been transverse to each other, so that the two forces partially neutralize each other in their action upon the selenium. Lastly, the current flows through not only the surface layer which is acted upon the light, but also through the portion which is underneath and not affected thereby, and which therefore detracts from the actual effect of the light upon the selenium at the surface.

Charles Fritts in 'On the Fritts Selenium Cells and Batteries' from the *Journal of the Franklin Institute* (1885)

We saw in chapter 20 how Einstein was able to use the existence of the photoelectric effect, where light blasts electrons out of metal to produce an electric current, to establish the reality of photons (and win himself a Nobel Prize). This process would prove even more effective when using semiconductors, leading to both the ability to produce 'free' energy from the Sun's light and to transform photography.

Plants vs technology

Capturing energy from sunlight has always been essential for life on Earth. Leaving aside the way the Sun keeps our planet warm enough to survive (thanks, in part, to the greenhouse effect), much of the energy that feeds life originates in sunlight. Through the complex mechanisms behind photosynthesis, plants convert energy from photons into chemical energy, used first to enable the plant itself to survive and then transferred to animals as they eat plants.

Now, of course, we are doing the same kind of thing with photovoltaics – solar cells – and in some ways beating nature. It can be difficult to make a precise comparison, as the two approaches do very different things with the energy from photons. The approach that presents photosynthesis in its best light is to use electricity from solar cells to drive an electrochemical process. In that case, the technology is two to three times more efficient than a plant. If, though, we allow the solar cells to do what they are designed for and generate usable energy, they can be as much as 20 to 30 times more efficient. It helps that photovoltaics are far less fussy about the specific colours of light they can handle, as well as having a more efficient approach to energy conversion from photon to electron.

Solar cells use a semiconductor where the energy of an incoming photon is used to push an electron to a higher energy level, producing an electron/hole pair. Each part of the cell acts as a diode, ensuring that current will only flow in one direction. Over time a whole range of materials have been developed to make these photovoltaic devices more efficient, some based on straightforward silicon, others built into flexible film that can be used on curved surfaces.

Standard cells are typically around 20 to 25 per cent efficient (where efficiency reflects how much of the light energy is converted into electrical energy), though high-end cells are approaching the 50 per cent mark. Although film-based cells are less efficient – typically a few percentage points below standard cells, and nowhere near the 50 per cent mark – they are so much cheaper to make that they usually cost less per amount of energy generated (though they are also likely to have a shorter lifespan).

From the eye to the camera

Another way that we use photons from the Sun to produce electrical signals is in detecting these photons. This is a process that has happened naturally for millions of years. When a photon reaches a receptor in the retina at the back of an eye, it boosts the energy of an electron in a so-called photoreceptor molecule. The combined effect of many individual sensors is pulled together as an electrical signal that travels up the optic nerve into the brain. Of course, the light source these days is just as likely to be manufactured as solar – but the effect is the same.

A similar process is handled very differently in a digital camera. Where the camera generates an image (we'll see how in a moment), the eye does not. Instead, the electron-carried signals from the optic nerve are processed by different bits of the brain which deal with aspects such as motion detection, shape and pattern recognition, colours and so on. In the camera things are much simpler. In effect, the sensors behind the lens are electron counters.

Digital cameras were first invented by Kodak in 1975, though initially unwisely suppressed as a threat to their film business. Just like the gas companies faced with electric lighting, rather than leverage their access to the market, Kodak did not become a digital camera giant. As a result, the envisaged threat was realised, with the company eventually filing for Chapter 11 bankruptcy in the US. Although the brand name has continued, the company is now a shadow of its former self.

When light passes through a digital camera lens, on your phone for example, the photons hit a mosaic of coloured filters to deal with the primary colours of light: red, green and blue – this is needed as the sensors themselves are not colour sensitive. Photons then arrive at sensors. Early cameras used a mechanism called CCD (charge coupled device), which is still used for some specialist applications. This is split into an array of cells, each of which accumulates electrons, knocked out of a semiconductor by incoming photons. The voltage of each cell is then measured to produce the final image.

The more modern approach uses a CMOS (complementary metal oxide semiconductor) chip. This is a circuit featuring an array of light sensitive diodes and amplifiers that makes a continuous measurement of the intensity of incoming light hitting each diode. CMOS sensors are both quicker to operate and cheaper than CCDs, so tend to dominate everyday use, though CCDs are still used where the highest quality of video is required, as CCDs build up a full picture, where CMOS cameras tend to build a line at a time, resulting in less quality with a changing image.

As Richard Feynman pointed out, the interaction of electrons and photons is universally important – here in the transfer of energy from light into electricity. We also saw in chapter 12 how the reverse can happen, transforming electrical energy into light. But the relationship between the electron and the photon had one more trick up its sleeve, inspired by another of Einstein's ideas: stimulated emission of radiation.

GETTING STIMULATED 33

Bond: Do you expect me to talk?
Goldfinger: No, Mr Bond, I expect you to die!
Dialogue from the film *Goldfinger* (1964)

We met light emitting diodes way back in chapter 12 – their practical development occurred in parallel with their more glamorous big brother, the laser. The concept for this was driven by an idea dreamed up in 1916 by none other than Albert Einstein, based on the new understanding at the time of nature of electrons in atomic structure.

As we have seen, light energy can both push an electron in an atom up to a higher orbital and be produced when an electron drops down a level. The majority of sources of light are the result of an electron being stimulated (for example, by heat in a flame) and then dropping back to a lower level. Einstein speculated that with the right technology it might be possible to use this process as a light amplifier.

Amplifying light

The idea would be that a photon of light would first push an electron up to a relatively unstable energy level. Then, before the electron dropped back to release a similar photon, that same electron might be hit by another photon. The doubly energised electron would then drop back to produce not one, but two photons. To expand the acronym 'laser', this would be light amplification by the stimulated emission of radiation.

It might sound as if light amplification is a relatively trivial thing – making a brighter light bulb. But the significance of the laser is much greater than this because it produces a different kind of light to a conventional light source. Each photon of light has a property known as its phase. This is something that changes direction over time – you can think of the phase as a fast-rotating clock hand attached to a photon. Phase in a photon corresponds to the position in a wave's cycle when we think of light as a wave.

Light from the Sun, a flame or a conventional light bulb consists of a stream of photons with a whole mix of phases. These influence how the beam will be dispersed when it interacts with atoms. However, the way that laser light is produced means that the photons' phases are in step – from a wave viewpoint, the waves are moving from side to side in unison as the light progresses. This 'coherence' means that laser light does not disperse in the same way as conventional light: as a result, a tightly focused, high-energy laser beam can feed so much energy to a material it touches that it can cut into it like a knife.

The stimulated emission process was first achieved in practice in 1954 by Russian physicists Alexander Prokhorov and Nikolay Basov, who had published a paper on using this mechanism to produce a device that amplified microwaves – still electromagnetic radiation, but of lower energy than visible light. The same year as Prokhorov and Basov published their paper, American physicist Charles Townes independently produced such a device, based on stimulating emission of radiation in a ruby crystal, which he named a maser.

Masers were interesting, and useful in the communications field where Townes worked, but the real prize was to get the same phenomenon working for visible light. A number of labs around the world were chasing this prize, with Art Schawlow, working with his former boss Townes at AT&T's Bell Labs (yes, there again), among the leaders of the pack, along with a former graduate student of Townes at Columbia University called Gordon Gould, who worked at defence contractor TRG.

Big money and politics

By the end of 1958 the race to produce a laser was hotting up. Townes and Schawlow had filed a patent for what they dubbed an optical maser, while Gould (who came up with the term laser) and TRG had asked for a $300,000 grant from the US government's Advanced Research Projects Agency. Such was the effectiveness of Gould's pitch, which included the promise to be able to send a communication to Mars and punch a hole through sheets of metal, they ended up being awarded nearly a million dollars. Gould gained a brief edge over his competitors when he realised that by putting mirrors at either end of the lasing material, it would allow repeated passes, building the excitation.

It was at this point that developments became farcical. Because they were working on a defence contract, Gould and his co-workers had to undergo security clearance. This was in an America that was under the sway of the House Un-American Activities Committee, deeply suspicious of any hint of communism. Gould, born in New York in 1920, had worked on the Manhattan Project towards the end of the Second World War, but was considered a risk because he had dabbled with communism.

The left-wing politics of his youth would continue to haunt Gould. In 1959, these were brought up when Gould's security clearance was assessed for working on his laser project for TRG. When his youthful political views were combined with the fact that he and his wife had lived together before marriage, and the problem that two of Gould's referees for security clearance had beards (which apparently made them potential subversives), his clearance was turned down.

Not only was Gould not allowed in his own lab, but he also was not even allowed to read his own notebooks, as his work was classified and he didn't have clearance. Meanwhile, his competitors at Bell Labs had their own problems. They had toyed with using a solid-state material for the laser, such as the synthetic rubies used in masers. But these had been dismissed as too inefficient at visible light frequencies. Instead, they were concentrating on using gases and metal vapours, which theory suggested were more likely to work, but were trickier to handle. This dumped on Schawlow a whole host of practical issues to overcome.

In the meantime, an American electrical engineer with a physics PhD working at the Hughes Aircraft Corporation was also undertaking a smaller-scale attempt on the laser. Born in Los Angeles in 1927, Theodore Maiman's combination of electrical engineering and physics training provided an ideal mix of the practical and the theoretical. He had already worked with rubies on a maser project and felt that it should be possible to also use them in a laser. He was suspicious of the calculations that Schawlow had used to show that rubies would be too inefficient to work – and experimentally discovered that rubies were about 70 times better than Schawlow had thought.

Maiman's more practically driven approach, as opposed to the academically minded attempts coming from the other labs, made him concentrate on what was realistically achievable. He avoided extremely corrosive high-temperature alkali metal gasses such as sodium and potassium, used in other labs at the time. Similarly, he kept away from cryogenically cooled solutions, which might have worked in a lab, but were totally impractical for most real-world applications. And rather than accept that, as many others argued, it would be impossible to use a ruby as the lasing material, he went ahead and proved them all wrong.

How to stimulate a ruby

This didn't mean that Maiman had everything cracked. To stimulate the ruby he needed a very bright light source, but the obvious solution, arc lamps, were so hot that they destroyed the fragile crystal. By luck, Maiman's assistant, Charlie Asawa, had a friend who was into photography and had recently bought a new product, an electronic flash, which was just starting to replace the old one-shot flashbulbs, based on burning out a strip of magnesium or zirconium in a high-oxygen atmosphere. Maiman managed to adapt a spiral flash tube from a flashgun to fit around a cylindrical ruby with mirrored ends. On 16 May 1960, Maiman produced the first series of laser pulses from his device.* The down-to-earth Maiman was horrified to

* One of the reasons Maiman was able to get a laser going first is that most of his competitors, who came from a communications background and were only thinking of lasers as a communication medium, assumed a laser had to produce a continuous beam, rather than high intensity pulses.

see the newspaper headlines that ensued, including: 'L.A. man discovers science fiction death ray.'

Writing in later years, Maiman can come across as bitter – but with good reason. He pointed out that ten people were awarded Nobel Prizes linked to lasers ... but the Nobel committee totally ignored the man who built the first laser – a bizarre decision. Almost as soon as Maiman's employer, the Hughes Corporation, announced the successful development of the laser, competitors who were more part of the academic establishment, apparently sneering at Maiman's work in industry, tried to minimise his achievement and to suggest that others really 'invented' the laser, even though their designs were never made to work. This counter-spin went on for decades.

Lasers rapidly became the big new thing, not only replacing masers in communications, but finding wider uses because their sheer concentration of in-step 'coherent' light made them potentially extremely powerful. It was only four years after Maiman's first successful test that James Bond (in the film *Goldfinger*) faced the prospect of being sliced in half by a (fake) laser directly based on a picture of Maiman's prototype. Over time, many different types of laser would be developed – the approach that became most commonplace was the use of electroluminescence as featured in the LED (see chapter 12), producing the laser effect from a solid-state semiconductor chip.

Lasers are pure quantum devices, dependent on our understanding of the way that electrons behave. By the time lasers were developed, it might seem that the electron had been thoroughly understood. Yet what might seem like simple enough questions remained unanswered – some still a mystery to this day. And one of the greatest of such questions was also, in a sense, the smallest. How big is an electron?

SMALL AND BEAUTIFULLY FORMED 34

To a request to explain what an electron really is supposed to be we can only answer, 'It is part of the ABC of physics.' The external world of physics has thus become a world of shadows. In removing our illusions we have removed the substance, for indeed we have seen that substance is one of the greatest of our illusions ... that physical science is concerned with a world of shadows is one of the most significant of recent advances.

Arthur Eddington in *The Nature of the Physical World* (1928)

As the decades have passed in our biography, in some ways we have uncovered a clearer picture of the electron. With first the establishment of it as a particle, then as a quantum phenomenon with wave/particle duality, we have seen our understanding grow, making it possible to not just be aware of the electron's role in electricity and the atom, but also to explicitly make use of it in our array of electronic technology.

A Goldilocks universe

Yet, the electron remains, as suggested by Eddington's evocative words in the quote above, an inhabitant of the world of shadows. Think, for instance, of two key aspects of the electron: its electrical charge and its nature as a particle. We don't know why the electron has the charge it does. More specifically, it is in some ways quite remarkable that the electrical charge on the three quarks inside a proton happen to add up to produce the exact positive equivalent of the negative charge on the electron. Without that match atoms would not exist.

I have seen it argued that they must have this charge or there would not be a consistent theory – but we can hardly expect physical reality to depend on whether or not it supports a human-originated theory. Perhaps all we can do is fit it into the bigger picture of the fine-tuning problem. This is that observation that many of the constants of nature seem to be fine-tuned for the existence of life. We live in a Goldilocks universe – if these constants had been slightly different, life (indeed, in many cases, matter) could not exist.

There are many such examples, but since our main character is a particle, just one example is that if neutrons (or, more precisely, the quarks that make them up) were 1 per cent lighter, atoms could not form. This is because all the protons would decay into a neutron, a positron and a neutrino. No protons, no atoms, no life. The charge match of protons and electrons is yet another aspect of this fine tuning without which atoms and life would be non-starters.

We have a good measure of the size of that charge, starting with Millikan's work. And we also know the mass of the electron to considerable accuracy. But what you might imagine is another fundamental aspect of electrons' nature remains far more in the shadows. How big are they?

So small it's not there?

The electron is generally thought of as a point particle with no size at all. As we have seen, there is a danger that such a value will produce theoretical effects that tend to infinity. It's important to separate what it's convenient to think to be the case and what appears to be the actual situation. Note size here is different from its nature as a quantum particle.

As we have seen, quantum particles like electrons or protons cannot be said to have a specific location, except when they interact with their environment and for an instant are specifically pinned down. If we observe the location of a free particle, we can envisage it spreading out as a growing sphere of probability, which has an effective size. Similarly, electrons in orbitals of an atom occupy a sizeable (compared with the particle) shape in space. But that isn't a meaningful definition of the size of the electron. What we are describing is simply the probability of it being at any particular location within that volume.

However, with other quantum particles, this nature doesn't prevent us from pinning down a size. Let's take the case of the proton and see how this was measured and what this tells us about electrons. The measurement was first made in 1956 by American physicists Robert Hofstadter and Robert McAllister. (By now it's no surprise that Hofstadter won the 1961 Nobel Prize in Physics for this work, or, perhaps, given the arbitrary nature of these awards, that McAllister didn't.)

They fired high-energy electrons at hydrogen gas. The nucleus of a hydrogen atom is just a proton, so this was a useful way of working with protons. If you imagine a macro-scale version of this experiment, where we might fire table tennis balls at an invisible football, we could learn something of the size of the football from the directions at which the smaller balls fly off. It's rather more complicated when firing electrons at hydrogen, both because each atom also contains an electron, and because there will be an effect from attraction between the proton and the incoming electrons. But with high enough energy in the electron beam and allowance for electromagnetic effects it proved possible to make a calculation.

Hofstadter and McAllister worked out that a proton has a radius of around 10^{-15} metres (a femtometre).* That's 1 divided by 1 followed by 15 zeroes. Over time, the measurement of the radius was tightened up to 0.877 femtometres. More recently, there have been

* Hofstadter referred to a femtometre as a 'fermi' in honour of the Italian physicist of that name, but though that term is used by some physicists, femtometre (with the same symbol of fm) is the standard name for this measurement.

suspicions that it is a little smaller. This was the result of measurements made in 2010 and later, using artificial atoms known as muonic hydrogen.

Hydrogen, as we have seen, is made of a proton with a single electron. Muonic hydrogen replaces the electron with a muon. This is another fundamental particle that is very similar to an electron, but has a much higher mass. Such 'atoms' don't exist in nature and rapidly decay, but can be created for experimental purposes. With its extra mass, the muon's orbital comes over 200 times closer to the proton, and makes for a more accurate measurement of the size of the proton based in the location of the orbital. The muonic hydrogen measurement came up with a radius of 0.842 femtometres. As yet, the exact value has not been resolved.

However, when similar approaches to those used to find the size of a proton are applied to an electron, scattering electrons with other electrons, for example, no size is indicated. At the very least it is too small to measure, and may have no value whatsoever. For various non-quantum calculations, the electron is assigned a 'classical electron radius', which is calculated by imagining the size of a charged sphere with the same rest mass energy as an electron – but this value is ridiculously large (bigger than a proton) and were it the electron's true physical size it would produce all sorts of anomalies.

All we can definitively say is that an electron's radius is no bigger than 10^{-18} metres – less than a thousandth of the size of a proton. This can be calculated as the largest possible size using the uncertainty principle of quantum theory. Best attempts at measurement using a mechanism called a Penning trap, used to isolate single charged particles, suggest it is no bigger than 10^{-22} metres. Of course, this in no way prevents it from being a true point with no size at all.

More chapters still to come

Our physicists have become so good at making measurements and developing theories that it can be easy to think we are close to pinning everything down about the electron. But this is very similar to what physicists thought was the case at the end of the nineteenth century in a different area of study.

In chapter 20 we discovered how German physicist Max Planck made the first tentative steps towards quantum theory, overthrowing the 'classical' physics of the nineteenth century. The young Planck had been an excellent pianist and could easily have become a professional musician. His physics professor at the University of Munich claimed that physics was a 'complete' science – all that a physicist could hope to do was make more accurate measurements. Thankfully, Planck wasn't put off enough by this apparent lack of opportunity to abandon physics for a musical career.

We have reached the present day in the biography of the electron – but we certainly don't know everything, and there are certainly more chapters to come. As I was finishing this book in late 2025, new research suggested an explanation for the behaviour of electrons that escape from solid matter, which had previously not behaved as expected. We have seen a number of ways electrons might be prised loose from a solid, whether it be due to the photoelectric effect or the impact of another electron. Yet for such electrons, it is not enough simply to get the necessary energy to break free.

Physicists at TU Wien have likened the situation to a frog trying to get out of a special box. The box only has a small window, partway up one side, through which the frog can escape. It's not enough that the frog has enough energy to reach the window – it also has to be on the correct trajectory to take it through the opening. A similar limitation has been found with solids, where different materials do not respond the same way to giving electrons enough energy to escape. We have seen how electrons in atoms can have different energy levels – it seems that not all levels with high enough energy to allow an escape are what the Wien physicists call 'doorway states' allowing a successful getaway.

Another example from 2025 was an experiment at the SLAC (Stanford Linear Accelerator Center) National Accelerator Laboratory at Menlo Park in California. This used an X-ray laser to be able to track the movement of a single electron in a molecule during a chemical reaction. Just as J.J. Thomson's experimental work made it possible to establish the existence of the electron as a particle, this experiment involved a new level of hands-on observation of electrons in action, giving a more accurate understanding than ever before possible.

We have seen just how much the electron does for us, how central it is to our electric lives, our electricity-powered environment and our quantum-based electronic technology. Yet if there is one thing this story suggests, it is likely there is more to come. The electron has been at the heart of reality since shortly after the Big Bang, 13.8 billion years ago. Over the last century or so it has matured with great rapidity. And it has plenty more to offer.

Watch this space.

FURTHER READING

In exploring the life and times of the electron, many detailed developments have to be passed over relatively quickly. If you would like to dive into more detail on particular aspects, here are some recommendations.

Atomic nucleus
The Fly in the Cathedral, Brian Cathcart (Viking, 2004) – in building up to the splitting of the nucleus, Cathcart gives an excellent picture of the gradual understanding of atomic structure.

Atoms
Atom, Piers Bizony (Icon Books, 2007) – excellent historical context on the discovery of the structure of the atom, but limited on science.
The Quantum Zoo, Marcus Chown (Joseph Henry Press, 2005) – better on the science, but not as good on the history.

Computing history
The Universal Machine, Ian Watson (Copernicus, 2012) – despite some editing issues, an approachable history of computing from the early steampunk days.

Paul Dirac
The Strangest Man, Graham Farmelo (Basic Books, 2009) – a long but impressively detailed scientific biography of this remarkable physicist.

Thomas Edison
Edison, Edmund Morris (Random House, 2019) – gives a much wider view of Edison's life than many biographies, though not particularly strong on the science.

Albert Einstein
Einstein and the Quantum, Douglas Stone (Princeton University Press, 2013) – there are vast numbers of books on Einstein's life and work, but the best explores his relationship with quantum physics, and hence with the electron.

Electric telegraph murder
The Peculiar Case of the Electric Constable, Carol Baxter (Oneworld, 2013) – enjoyable coverage of murderer John Tawell's capture using the fledgling electric telegraph.

Electrical charge
Charge, Frank Close (OUP, 2024) – an exploration of the nature of electrical charge and the impact of its existence on matter being neutral.

Electron market
Life's Engines, Paul Falkowski (Princeton University Press, 2015) – shows how microbes are central to the electron market that makes life possible.

Michael Faraday
The Electric Life of Michael Faraday, Alan Hirschfeld (Walker, 2006) – a detailed look at Faraday's scientific work, particularly concerning electricity.

Fine tuning problem
The Multiverse, Brian Clegg (Icon Books, 2025) – gives more detail on the fine-tuning problem and how attempts to explain it away with multiple universes having different constants of nature involve a logical fallacy.

Benjamin Franklin

Benjamin Franklin: An American Life, Walter Isaacson (Simon & Schuster, 2003) – Franklin has surprisingly few good biographies, but Isaacson gives arguably the best balance of science and biography.

Luigi Galvani

Galvani's Spark, Alan McComas (OUP, 2011) – heavy duty history of science (and very pricey), but impressive detail.

The Lunar Society

The Lunar Men, Jenny Uglow (Faber & Faber, 2002) – A straightforward, if fairly light, introduction to the eighteenth-century Midlands scientific society.

Guglielmo Marconi

Signor Marconi's Magic Box, Gavin Weightman (HarperCollins, 2003) – a real page turner in the story of Marconi's development of wireless technology.

James Clerk Maxwell

Professor Maxwell's Duplicitous Demon, Brian Clegg (Icon, 2019) – a detailed look at the life and work of the man behind the Cavendish Laboratory and the theory of electromagnetism.

Newton's Principia

The Principia: A New Translation, I. Bernard Cohen and Anne Whitman (University of California Press, 1999) – if you want to take on Newton's imposing masterpiece, this edition has an excellent, if challenging, guide to its meaning.

Periodic Table

The Periodic Table: Its Story and Its Significance, Eric Scerri (OUP, 2019) – Scerri is a leading authority on the periodic table and gives it in-depth coverage. Not a light book, but definitive.

Quantum Electrodynamics
QED: The strange theory of light and matter, Richard Feynman (Penguin, 1985) – not always an easy read, but a key introduction to this crucial quantum theory.

Quantum Technology
The Quantum Age, Brian Clegg (Icon Books, 2014) – a look at how quantum physics and our understanding of the electron have shaped the technology of modern life.

Radioactive products
Half Lives, Lucy Jane Santos (Icon Books, 2020) – fascinating look at the use of radium from its discovery to incorporation into a surprising range of products.

Spiritualism and Scientists
Ghosted, Alice Vernon (Bloomsbury, 2025) – reveals how a number of physicists, notably William Crookes and Oliver Lodge, were taken in by mediums.

Superconductivity
Ten Days in Physics that Shook the World, Brian Clegg (Icon Books, 2021) – 'Saturday, 8 April 1911' chapter offers a detailed look at Kamerlingh Onnes' day of discovery.

Telephony
The Telephone Gambit, Seth Shulman (W.W. Norton, 2008) – interesting exploration of just who did invent the telephone, based on archive material.

Nikola Tesla
Nikola Tesla and the Electric Future, Iwan Rhys Morus (Icon Books, 2019) – there are plenty of biographies of Tesla, most of which focus on his personality and totally exaggerate his scientific contribution. This is the best in giving an accurate picture of his electrical work.

J.J. and George Thomson
A History of the Electron, Jaume Navarro (Cambridge University Press, 2012) – academic biographical context on the Thomsons and their work on the electron.

Time Travel
How to Build a Time Machine (UK: Build Your Own Time Machine), Brian Clegg (St Martin's Press, 2011) – all you need to know about the physics of time travel, including time-travelling electrons.

Transistors
Ten Days in Physics that Shook the World, Brian Clegg (Icon Books, 2021) – 'Tuesday, 16 December 1947' chapter explores the events leading up to the day of the first transistor test.

INDEX